THEY LEFT THEIR MARK

WILLIAM AUSTIN BURT AND HIS SONS, SURVEYORS OF THE PUBLIC DOMAIN

THEY LEFT THEIR MARK

WILLIAM AUSTIN BURT AND HIS SONS, SURVEYORS OF THE PUBLIC DOMAIN

by

John S. Burt

LANDMARK ENTERPRISES
10324 Newton Way
Rancho Cordova CA 95670

ISBN 0-910845-31-X

Printed and typeset in 1986 in United States of America by
Edwards Brothers - Ann Arbor, Michigan
Cover design by Kathryn Ross
Book Design by Roy Minnick

TO THE PROFESSIONAL SURVEYOR

TABLE OF CONTENTS

FORWARD by John D. Voelker   xv

INTRODUCTION AND ACKNOWLEDGEMENTS   xvii

PART ONE

I.   IN THE BEGINNING

    Massachusetts   1
    Early Years in New York   3
    Wales Center, New York   4
    Journey to the West   5
    Preparation   6

II. OPPORTUNITY IN MICHIGAN TERRITORY

    Settlement   9
    Politics   11
    Battle Creek   11
    Masons   12

III. AMERICA'S FIRST TYPEWRITER

    Burt's Writing Machine   13

IV. A NEW CAREER

    Preparation   23
    U.S. Public Land Surveys   24
    W.A. Burt's First Government Survey   26

V.   QUEST FOR ACCURACY

    Burt's Second Government Survey   28
    Variation of the Compass   31
    The Solution   31

VI. TRIAL AND ERROR

    Iowa Surveys   35
    Burt's Variation Compass   37
    Difficulties   38
    Michigan's Internal Improvements   40

PART TWO

VII.   ACQUIRING THE UPPER PENINSULA

        Michigan Statehood   43
        More Survey's Are Needed   45

VIII.   THE UPPER PENINSULA SURVEYS BEGIN

        Burt's Contract   47
        Burt's Solar Compass   50
        Burt's Sun Dial   51

IX.   EXTENDING THE UPPER PENINSULA SURVEYS

        New Assignment   52
        Douglass Houghton   53
        Fraudulent Surveys   55

X.   A MOUNTAIN OF IRON

        Burt's Solar Compass   56
        Alvin Moves to Iowa   57
        Discovery   57

XI.   1845 AND 1846

        Legend of The Pine Stump   63
        Continuing the Upper Peninsula Surveys   65
        More Discoveries   66
        The Burt's Chalets   67
        Alvin   67
        The Solar Compass   70

XII.   1847 AND 1848

        The Michigan-Wisconsin Boundary Survey   75
        Closing the Township Lines   77
        Lucius Lyon's Concern   79
        Improving the Solar Compass   80

XIII.   1849

        Political Changes   81
        A Matter of Conscience   81
        Washington D.C. in December 1849   84

PART THREE

XIV.   JUSTICE FOR ALL

      Precedent   91
      Burt's Astronomical Compass   92
      Rectifying a Misconception   93
      Pending Litigation   94
      The Witness Tree   95
      Solar Compass and the Instructions   96
      The Iowa-Missouri Boundary   96
      Burt's Petition   98
      Preparations for the Great Exposition   99

XV.   NEW FRONTIERS

      The Great Exposition of 1851   100
      The Solar Compass and the Opening of the West   104

XVI.   JOHN BURT IN THE UPPER PENINSULA

      John's Bold Prediction   107
      Lobbying for the Soo   109

XVII.   1852 TO 1855

      Confirmation on the Iowa-Minnesota Boundary   111
      Burt's Petition   112
      Preparations for the Soo Canal   113
      Michigan's Legislasture of 1853   113
      The First Taste of Victory   114
      Railroad to the Upper Penninsula Iron Mines   115
      The Peninsula Iron Company   117
      The 33rd Congress   117

Table of Contents (Continued)

PART FOUR

XVIII.   1855

    Completion of the Soo Canal   121
    The Lake Superior Mining Journal   123
    The Solar Compass is Challenged   124
    Burt & Bailey   124
    The Manual of Instructions   125
    William A. Burt, Author   126

XIX.   1856

    The Equatorial Sextant   129
    The Compensation Debate Continues   130

XX.   1857 and 1858

    John at the Soo   133
    Completing the Upper Peninsula Railroad   135
    Time Runs Out   135

XXI.   THE BURTS IN THE UPPER PENINSULA

    Marquette: Sparks of Growth and Disaster   139
    John Burt - The Politician   143
    John Burt - The Inventor   143

XXII.   THE FINAL CHAPTER

    The Burts vs. the U.S. Government   147
    The Typewriter   149
    Burt's Equatorial Sextant   149
    The Solar Compass   150
    The Marquette Iron Range   152
    Evidence of the Marks   152
    Other Recognition   154

REFERENCE NOTES   155
APPENDICES
    A. The Burt Lineage   175
    B. The Author's Connection   176
    C. The Burts' Surveying Contracts   177
    D. The Burts' Iron Mining Business   182
    E. Burt's Solar Compass   183

INDEX   184

LIST OF PLATES

1.   The Burt crest   2

2.   George Washington as surveyor   3

3.   Map of the Northwest Territory - 1806   5

4.   Willaim Austin Burt   10

5.   John P. Sheldon (1792 - 1871)   15

6.   Burt's Typographer - America's first typewriter   16

7.   Burt's Typographer - top view   17

8.   Demonstration of William A. Burt's typographer   18

9.   Restored patent office drawing   18

10. Letters Patent - July 23, 1829   19

11. Burt's letter to his wife Phebe   20

12. Lucius Lyon (1800 - 1851)   24

13. The Rectangular Rystem of land division   25

14. Abraham Lincoln, surveyor   30

15. Reproduction of Burt's original "variation compass" - patented 1836 32

16. The Fifth Principal Meridian - Iowa Territory   36

17. Burt's patent solar compass made by R. Whitcomb   39

18. The Toledo Strip   44

19. The northern point of the Michigan Meridian as established in 1840   48

20. Burt's improved solar compass made by William Young   49

21. Burt's Sun Dial   51

22.   Dr. Douglass Houghton (1809 - 1845)   53

23.   Burt's survey party camped on the edge of Teal Lake   59

24.   Burt's field notes from east boundary of
      Township 47 North, Range 27 West   61

25.   William Ives' map of Iron Hills   62

26.   Negaunee seal   64

27.   William Austin Burt geological map - 1845   68

28.   William Austin Burt's geological map - 1846   69

29.a William A. Burt's chalet - 1909   70

29.b William A. Burt's chalet - 1982   71

30.a Chalet of Wells Burt - 1902   72

30.b Chalet of William Burt - 1902   73

30.c Chalet of Austin Burt - 1902   73

31.   Assumed Michigan-Wisconsin boundary - 1837   74

32.   Tamarack tree marked by William A. Burt on June 7, 1847   76

33.   Completed Michigan-Wisconsin boundary line   78

34.   Aluminum Burt Solar compass (c. 1880)   80

35.   William Austin Burt - face and signature   83

36.   U.S.Senator Alpheus Felch   86

37.   U.S.Senator Lewis Cass   87

38.   J. Bailey advertisement   93

39.   Witness tree marked by William A. Burt - June 17, 1850   95

40.   Fictitious township - General Instructions of
      1850 (Ohio, Indiana, Michigan)   97

41.	The Crystal Palace	100
42.	American exhibits at the Great Exhibition	102
43.	A U.S. Deputy Surveyor operatiing a Burt solar compass	104
44.	California Surveyor General's seal depicting Burt's solar compass (1885)	105
45.	John Burt	106
46.	Marquette, near the Carp River - 1851	108
47.	Lake Superior Iron Mine No. 1	116
48.	Peninsular Iron Company's Magnum kiln	118
49.	Early construction of the lock pit for the Soo Canal	122
50.	Completed Soo locks of 1855	123
51.	Burt & Bailey advertising card	124
52.	Burt's solar compass made by John Bailey in 1852	125
53.	Burt's Equatorial Sextant - 1856	128
54.	Soo Canal lock (1870's)	134
55.	Phebe Burt and Willaim Burt	136
56.	Burt Monument	137
57.	Marquette in 1881	138
58.a	Wells Burt	140
58.b	Austin Burt	140
59.	John Burt house	141
60.	Burt Block in Marquette	142

List of Plates (Continued)

61.	Burt Brothers business card	143
62.	Ore Dock in Marquette (1872)	144
63.	John's lock patent drawing	145
64.	Weitzel Lock nearing completion in 1881.	145
65.	Last bill for relief of Burt's heirs	149
66.	Surveyor's chain (1875)	151
67.	1878 Burt solar compass made by W.& L.E. Gurley	151
68.	Burt Lake plat map - 1841	153

Endpapers

Map of portion of the Marquette Iron Range showing
Township 47 North, Range 27 West taken from
Sketch of the Public Surveys in Michigan, issued by the
Surveyor Generals Office, dated November 8, 1854

Foreword

The story of how the Upper Peninsula finally became a part of Michigan must have made the angels weep. And doubtless also giggle. Similarities in climate and geology would have given it to Canada; simple geography and sheer propinquity would have meant Wisconsin; but the vagaries of politics and the pressing need for compromise instead landed it in the reluctant lap of a nose-thumbing Michigan way down there below the Straits. Take a look at the map.

The then territory of Michigan, preening itself over the prospects of statehood, had instead coveted a narrow strip of land along the Ohio border known as the Toledo strip. The place was already settled and measured (the survey lines, interestingly enough, having been run by a recently-graduated army cadet called Robert E. Lee) and inhabited by prosperous farmers and other citizens who regularly paid their taxes. Above all it was not an unexplored wilderness like the U.P., consisting mostly of rocks and swamps and bands of roving Indians.

But the already admitted state of Ohio had thought otherwise. Tempers had flared and people had begun gathering along the disputed border. Soon there were rumors of shots being fired in what some began calling the "Toledo War." So Congress had met and huffed and puffed and—guess what?—awarded the disputed strip to the state that already had votes in Washington, that is to Ohio, while poor territorial Michigan had to settle for the scorned U.P.

As the still rankling Michigan began hearing rumors of copper and other minerals being found in the U.P., it sighed and shrugged and started a modest geological exploration of the area. When, in 1836, it finally became a state (with itself now having votes in Washington) the Government relented enough to listen to the arguments of a young Douglass Houghton that it merge its ongoing national linear survey with Michigan's modest mineral search, paying most of the cost, and appoint him the leader of the expedition.

And this is how it happened that the evening of Sept. 18th, 1844 found Houghton's chief surveyor William Burt and his survey crew camped out along the east shore of Teal Lake in the Upper Peninsula within scarcely a mile of the next day's big discovery—a veritable mountain of rich iron ore.

This is not the place to go into the details of that discovery or of how the magnetic compasses spun and became useless as Burt and his crew drew near the mountain of ore, and that only the solar compass really told them where they were. Enough to say that on that day they discovered two things: a mountain of ore that had lain there neglected for unknown millions of years and that its site was henceforth to be known as Township 47 North, Range 27 West of the Michigan Meridian.

Burt was a 52-year-old self taught Yankee land surveyor when his discovery was made. Fittingly enough his story is told here by one of his direct descendants five generations removed. And, ironically enough, the man seems not to have made a dime out of his discovery or have even tried to. Instead he went on surveying and creating other ways for other men to make money and develop ulcers. Some idea of the wealth that this man helped to unlock for the state that didn't want the U.P. is that a recent source estimates that the value of iron and copper so far shipped from the U.P. exceeds that of all the gold ever mined in California. No wonder the angels giggled.

I first heard of William Burt when I wrote my only historical novel called *"Laughing Whitefish."* It was the story of a Michigan law suit growing out of Burt's find and brought by the surviving daughter of a Chippewa Indian chief who knew one of the Indian backpackers who was with Burt the day of the big discovery. Seems some later downstate entrepreneurs, hearing rumors of Burt's find, had tracked down the Indian chief and promised him a pittance if he'd tell them where. The trusting Indian had told; white man had prospered; and—again guess what?—had somehow forgotten to pay.

Back then not too much was known about the story of William Burt, though one of the best sources I found was an excellent book by John Bartlow Martin about the U.P. called *"Call It North Country."* One thing I learned is that William Burt preferred to roam and measure the land not for money but because he loved it. With this biography I am also learning that William Burt was probably one of the best of our neglected native geniuses.

John D. Voelker
(Robert Traver)

Introduction and Acknowledgements

It was June 1840 when U.S. surveyor William Austin Burt first entered Michigan's rugged Upper Peninsula, a wilderness that, according to Henry Clay, stretched "beyond the remotest settlement of the United States, if not the moon." Neither Burt nor his sons who assisted him could have envisioned the role they would play in helping shape the future of this area acquired from Wisconsin Territory in 1837.

William A. Burt typifies the American pioneer as the rugged surveyor who explored and accurately charted the wilderness. He led the way for settlement and growth in Michigan and the midwest. Throughout his life Burt adhered to the strong religious and moral principles that formed the foundation of his strong character. His word was his bond. His integrity was never questioned. With inventive mind and a creative spirit he persevered, mainly for the benefit of others. By his lack of personal greed William Austin Burt left a lasting legacy of public service.

They Left Their Mark is the first full story of William A. Burt and his five sons who assist him on the public land surveys. It centers on William Austin who discovers a new source of wealth for Michigan and the nation, an event that confirms the value of Burt's unique invention for surveying the expanding U.S. frontier. By then he has already invented America's first typewriter. It also follows the activities of his son John who becomes a leader in the growth and development in Michigan's Upper Peninsula, particularly Marquette County.

As a great-great-great grandson of William A. Burt, I had heard about this distinguished ancestor when I was young. As a child, however, little of the information registered. It was during the euphoria of the preparations for the nation's bicentennial celebration that I began a search for more information about William Austin and his sons. It started in a dust-covered box in my garage containing a wealth of family records, and it culminates more than a decade later with the publication of *They Left Their Mark*.

I gratefully acknowledge the encouragement and assistance in this research of the Burt family, many surveyors, and reference personnel. John H. Clarke, a Burt descendant previously unknown to me, sent many items originally belonging to the early Burts, including books, photographs, furniture, and an 1841 Burt sun dial. A special family reunion was held in Carthage, Missouri, by the Deckers and the Wings, Burt descendants, in conjunction with our research trip to Michigan. The Strongs of Des Moines, also relatives, supplied information about Burt's work on the Fifth Principal Meridian.

Extensive family papers and a leather-bound first edition copy of Burt's *Key to the Solar Compass* were contributed by John Shields Burt of Palo Alto. Aunt Dorothy Burt Findlay, of Vista, California, provided several items from the family heirloom, including the only known daguerreotype ever taken of William Austin.

A debt of gratitude goes to my brother Richard A. Burt of San Diego, who helped in the editing of the entire manuscript. Jim Burt, my twin brother of Los Gatos, California, contributed valuable research and promotional help. Special love and appreciation go to my wife Carol and our two daughters, Karen and Kathy, whose patience and understanding throughout these many years have made this project possible.

The surveying profession has provided gratifying support to this project. Members of the Bureau of Land Management (BLM) who contributed useful information include U.S. Surveyor General Bernard W. Hostrop; Paul C. Herndon, Office of Public Affairs; C. Albert White, surveyor-author; Kenneth D. Witt, Chief Cadastral Survey, Denver Branch; and William W. Glenn, Chief, Cadastral Survey, Portland Branch. Glenn maintains the only supply of Burt solar compasses currently used by the BLM.

Members of the Surveyors Historical Society—particularly Directors Francois D. "Bud" Uzes and Cecil H. Hanson—have provided considerable material and advice. Surveyors Kenelm C. Winslow of El Prodo, New Mexico, and Stan Weaver of Los Angeles have contributed additional useful information.

At the 1985 convention of the Michigan Society of Registered Land Surveyors, I was given a useful manual entitled *The History of Land Surveying in Michigan*, authored by Ralph Moore Berry, P.E., R.L.S. The Society has taken a special interest in the project.

Richard L. Drahn, U.S. Forest Service surveyor, sent a detailed map showing the section lines run by the Burts in Upper Peninsula's Ottawa National Forest from 1845 through 1851. Robert C. Miller, of Pittsburgh, an expert on William J. Young, the first manufacturer of Burt Solar Compasses, has rendered discerning advice and supplied over 50 copies of correspondence between Burt and Young, from the Burt Papers, Marquette County Historical Society, Marquette, Michigan.

The historical significance of a complete biography of William A. Burt and his sons was impressed upon me by LeRoy Barnett, Reference Archivist of the Michigan Department of State, who sent copies of pertinent records from the Michigan State Archives. Professor Alan S. Brown of Western Michigan University in Kalamazoo, author of articles about William A. Burt, has provided considerable information about Burt and Michigan history.

In California the research libraries at U.C.L.A. and Claremont Colleges provided me with essential information about Michigan history and U.S. Government documents, including land office records. The assistance of Julia Jacinto and her staff at the latter facility is especially appreciated.

Additional General Land Office records on microfilm were obtained from the National Archives, Washington, D.C. Deborah Jean Warner, Curator, History of Physical Sciences, at the Smithsonian's National Musuem of American History, has sent pictures and material concerning William A. Burt's inventions, three of which are among the National Museum's collection.

The single most valuable research facility, however, has been the Marquette County Historical Society's Longyear Research Library which maintains the most complete collection of original Burt Papers. During two extended trips to Marquette, considerable courtesy and assistance were provided by Directors Ester Bystrom and Frank O. Paull, Jr., and their staffs. Burton H. Boyum, Board of Trustees President, has also been most helpful with historical information about the Upper Peninsula.

While researching in Marquette, noted author John D. Voelker (Robert Traver) graciously traveled from his home in Ishpeming to visit with us at the Historical Society museum and library. Voelker's historical novel *Laughing Whitefish* was a major source of inspiration for *They Left Their Mark*. Frank Matthews, colorful curator of the Jackson Mine Museum, Negaunee, Michigan, proudly displayed artifacts of the mining era and cordially shared local stories of the Upper Peninsula pioneers.

Finally, special gratitude goes to Leah Cannon Atwater, a descendant of George H. Cannon—surveyor and friend of William A. Burt. George Cannon married Burt's niece, Lucy Cole, and he authored several articles about William A. Burt. Mrs. Atwater became my contact at the University of Michigan and forwarded copies of significant material pertaining to William A. Burt and Lucius Lyon.

My heartfelt thanks to each of you who have made *They Left Their Mark* a reality. The completion of this project was possible only through your dedicated effort and gracious support. I cherish the many friendships that I have made during the past decade of research, and I am enriched with a unique appreciation for Michigan, the surveying profession, and my ancestors—William Austin Burt and his sons.

John S. Burt
Orange, California
May 1986

PART ONE

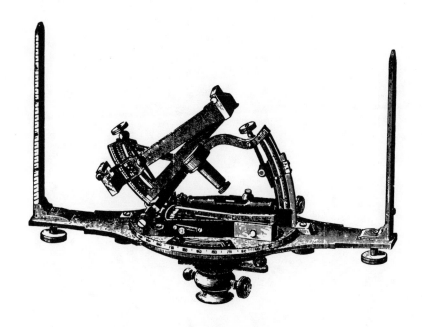

Chapter I

IN THE BEGINNING

It is indeed a desirable thing to be well descended,
but the glory belongs to our ancestors.

-translated from Plutarch   (A.D. 46 - 120)

Massachusetts

It was 1638 when the Richard Burts sailed from England to Taunton, Massachusetts.[1] Fishermen were attracted to this small village, then called Cohannet, by the herring runs that swarmed up the Cohannet and Taunton rivers to spawn. While fishing may have appealed to some of the sea-faring members of the family, most of the early Burts preferred farming.

Like many of the first settlers to the area, Richard Burt emigrated from Devon or Somerset, England. It is probable, however, that this first generation of American Burts came via Barbados where Richard is listed as an early landowner.[2]

In 1672 his son Richard, a surveyor of highways, helped lay out the road from Weir Village to Assonet Neck. He also served in Taunton as a local constable and was a father to eight children.

This was 113 years before the rectangular system of land surveying was established in America. The irregular English metes and bounds system used in the colonies worked just fine until disputes over property lines arose, not unexpectedly. An example of the problem is found in the will of the second Richard Burt, in which he granted his eldest son Abel property described as

> a certain tract of land lying on the east side of the Taunton River, extending from the said river eastward to the highway which goeth down to the farms betwixt the land of James Walker and the land of Thomas Burt.[3]

Abel, born in 1657, later operated a sawmill and tannery in Taunton and served as a deputy sheriff. With his brother Ephraim he bought a 15 mile long tract which Abel cultivated and improved until his death. It was during this period, in 1684, that the people of Taunton voiced one of the first Yankee protests against taxation without representation.[4] When the governor demanded poll and property taxes, the town clerk was sent to deliver the message that if they could not have a voice in the assembly they would pay no taxes. The governor promptly jailed the messenger for three months. When he was released, the grateful townspeople awarded him 100 acres of land in appreciation.

Abel, Jr., Abel's second son, became a lieutenant in the First Fort Company and was a Commissioner of Taunton. His son George carried on the tradition of farming, and was the last of this line of Burts to remain in the Taunton area. In 1757 George moved his family to the farm of his maternal grandfather, Thomas Briggs, where his son Alvin Burt was born in 1761.

When the American Revolution began, Alvin enlisted as a private and served in a number of regiments during 1777 and 1778. The following year he married Wealthy Austin, also of Taunton. By 1790 Alvin grew restless and moved his expanding family to the village of Petersham, about 60 miles northwest of Boston.

1.   Burt Coat of Arms *courtesy Jim Burt*

Families dependent on farming for their livelihood often raised many children, and the Burt family was no exception. William Austin Burt was born in Petersham, Massachusetts, on June 13, 1792, the fifth child of Alvin and Wealthy. Two of their sons, Alvin, 12, and George, 10, were then old enough to harvest the crops with their father. Son Zelotas, age 3, and daughter Wealthy, age 4, would keep their mother busy along with young William.

The name "William Austin" came from Wealthy's father, an English sea captain she hardly knew. She was just three when he sailed from Massachusetts with a cargo destined for the East Indies. Her 14-year old brother accompanied her father on the voyage, and both were lost in a severe storm at sea. The incident profoundly effected Wealthy and her son William Austin's life work.

Petersham had been incorporated almost 50 years earlier, in 1754, but it remained a small community. Schools were few and books were scarce. Most of William Austin's education was self taught, although he did attend some classes but just long enough to master the alphabet. He would later recall that unlike most other children his age he had an insatiable quest for knowledge. He chose to dig into books rather than dirt, and he found more enjoyment in the solitude of learning than the competition of team sports.

It was evident even at an early age that William Austin had an inquisitive mind. He was particularly fascinated with the rising and setting of the sun, the moon and its' curious smile, and with the stars that twinkled in the darkened sky. When he inquired about them, he was never quite satisfied with the simple answer that "God made them."[5]

Early Years in New York

Shortly after the turn of the century, in 1802, a severe crop failure forced Alvin to sell his Petersham farm. Like many New Englanders in search of opportunity he moved his family and worldly possessions westward, to a beautiful and fertile valley on the southern edge of the Catskills. There he established a small farm in Freehold, about 40 miles south of the large trading center of Albany.

By now Alvin's and Wealthy's family had grown larger, despite the loss of William Austin's older brother George, who died at the age of 17. Samuel was born in 1794 and Sally followed three years later.[6] Even with four strong boys to share the workload, there was always more work to do around the farm then daylight hours permitted. Consequently, William Austin acquired much of his early knowledge from borrowed books read at night by the light of a burning pine knot.

2.   Surveyor George Washington *(Lola Cazier, SURVEYS AND SURVEYORS OF THE PUBLIC DOMAIN, GPO)*

After an unsuccessful year in Freehold, Alvin and his family moved in with George Burt, William Austin's grandfather, who owned a farm in Broadalbin, New York, about 40 miles northwest of Schenectady. An abundance of lakes and streams surrounded the fertile land.

These were formative years for young William. Sadness occasionally intervened. Grandfather George passed away at age 78, and Zelotas, his teenage brother, drowned accidentally. The farm, however, continued to sustain the family, and they remained in Broadalbin for over seven years.

Fortunately, a neighbor who had been a college professor in Scotland owned a large personal library. From this valuable source Burt soon mastered mathematics and astronomy. He invented a device that held his book open so he could read while both hands were occupied making shingles.[7]

One of his favorite books was a navigation manual belonging to an uncle. The subject so fascinated him that he talked of someday becoming a ship's captain. But his mother insisted that he reconsider. She could not bear losing her son as she had lost her father and brother. At her insistence he turned his attention to surveying, a form of navigation on land.[8]

The profession of surveying had been held in high esteem since the days of Euclid. Great leaders, such as George Washington who laid out boundaries of many farms in Virginia, were attracted to it. From a surveying text Burt learned the traverse tables. His mechanical skills became evident when he built a quadrant without ever having seen one, and he accurately determined the location of his father's house.[9]

Not far from the Burt farm in Broadalbin lived the family of John and Sarah Cole. Sarah had lost a child in delivery about the time the Burts moved there.[10] It was the third time one of their children had died within the first month of life. John Cole had married Sarah Wells shortly after concluding his tour of duty in the Revolutionary War. Full of enthusiasm, the Coles built their home in Broadalbin. Despite their losses, they raised seven healthy children, including their eldest daughter Phebe. She was about the same age as William Austin Burt, and a strong attraction between William and Phebe had developed by 1810, when William was 17 and the Burt family moved to Wales Center, New York.

Wales Center, New York

The new Burt farm was located on the edge of Buffalo Creek near Hall's Hollow (Wales Center), 25 miles southeast of Buffalo. The western migration had taken the family nearly 300 miles from Broadalbin. At age 18, William Austin purchased a second-hand surveying compass and made surveys in his neighborhood. On the fourth of July, 1813, he and Phebe Cole were married. Shortly afterwards the rest of the Cole family moved to Wales Center and established a farm nearby.

At the time William Austin and Phebe were married the United States was at war with Great Britain. The war of 1812 was unpopular from the beginning, precipitated when the British fleet began restricting the profitable U.S. trade exports from New York and other states. The conflict began on June 19, 1812, and William Austin and his father enlisted shortly afterward. William's first 60 day tour included participation in an unsuccessful attack on Ft. Burlington, Canada. The U.S. militia was poorly trained and ill equipped. Discipline was lax and morale low. While American troops managed to burn York (Toronto), the British retaliated by burning Buffalo while the 500 residents, who had evacuated the city, watched helplessly.

On April 11, 1814, John Burt, the first son of William and Phebe, was born. The new father scarcely had time to adjust to his new role, for in July he enlisted as a fife major for a second 60-day tour. On Christmas eve the futile war ended. By then William Austin had embarked on a new business venture with his father-in-law, John Cole.

It was the first general store in town and Burt's first attempt as an entrepreneur.[10] Unfortunately, the war had driven up prices, and the cost to transport goods from Albany was prohibitive. The partnership failed, but it wasn't long before William Austin began another business venture. Burt & Allen, Millwrights, was formed with John Allen, a friend and excellent mechanic. They built several mills in Erie County, although the work was sporadic.

Before long Burt became well known in the community. He was appointed a justice of the peace, school inspector for Erie County, and town clerk for Wales Center. In July 1816 he received his Masonic degree.

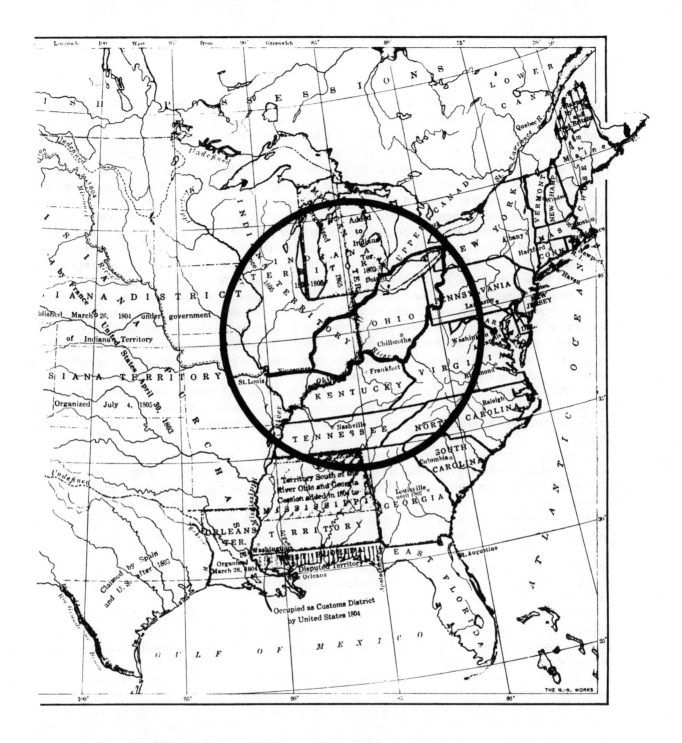

3. Map of the Northwest Territory - 1806 *from E.M. Avery, A HISTORY OF THE UNITED STATES AND ITS PEOPLE, Burrows, 1910, VII. 348*

Journey to the West

By 1817 William Austin had grown restless. Construction had begun on the Erie Canal, and new opportunities would soon be opening up in the West. At an early age he had resolved to choose a career that would utilize his skills for the betterment of mankind. His reward would be the self-satisfaction of providing the world with something of value.[11] At 25, however, his life seemed to be slipping away. Business was slow, and his responsibilities were increasing, particularly following the birth of his second son, Alvin, on June 19, 1816.

In mid-August 1817, Burt began a 10-week trip westward, as far as St. Louis. Choosing the southern route, he canoed down the Allegheny River from Lake Chautaqua to Pittsburgh, already a major manufacturing center for glassworks, steamboats, and other machinery. Continuing down the Ohio River to Cincinnati, he recorded in his diary that the villages were "handsome and healthy, and the country the best I ever saw."[12] Cincinnati was only a small village in 1817. The shops, he noted, looked just like those he had seen in Albany, while groceries and dry goods were priced about the same as in Buffalo.

As he neared St. Louis he recorded the soil conditions, much like a surveyor preparing his field notes: "flat prairie, good land, timbered with black walnut, oak, and maple." By canoe and on foot he averaged about 25 miles each day, arriving in St. Louis on September 19th. St. Louis was then just a small French village, acquired 14 years earlier from Napoleon as part of the Louisiana Purchase. The trip had been exhausting, but exciting. It had been five weeks since William Austin had seen his family. Following a brief visit to St. Charles, he headed home via the northeast route to New York.

In Vincennes, Indiana, soon after recuperating from a severe respiratory infection, Burt met two other young explorers, who accompanied him on to Detroit. Their journey covered several hundred miles of territory, uninhabited except for a few Indian villages and military forts. For several days a single chunk of venison was the only food they had to share. When they finally reached an Indian encampment their hunger was eased with a dinner of succotash and squash, purchased for a quarter. Burt recorded "the indians we met with appeared to be friendly and good humored, and the likeliest I ever saw."

Fort Wayne, built 23 years earlier at the junction of the St. Joseph and St. Mary's rivers, had been the site of only a few skirmishes with the Indians since General Anthony Wayne's victory in the battle of Falling Timbers on August 20, 1794. In fact, two years after Burt's visit the fort was closed from inactivity.

Continuing their journey, the three men reached Ft. Defiance in Ohio where Burt recorded, "poor land." He had confirmed what surveyor Benjamin Hough had found and reported two years earlier when he established the southern point of the Michigan Meridian. This was the watershed for the eastern and western flowing rivers, with an abundance of marshes and lakes. On one occasion Hough became so immobilized he had to wait for a hard frost to form to get out. It was Hough who informed Surveyor General Edward Tiffin in 1815 that the land was worthless.[13] On November 30, 1815, Tiffin then wrote the often quoted letter to General Land Office (GLO) Commissioner Meigs that the whole two million acres of bounty land in Michigan would not be worth the expense of surveying, nor would it contain one percent of usable land.[14] As a result, bounty lands were established in Illinois and Missouri instead of Michigan.

Although Tiffin's report led to a temporary halt in the land surveys in Michigan Territory, work on the Michigan Meridian was eventually resumed. As Burt passed through the area in 1817, however, he could hardly have envisioned that 23 years later he would be selected to establish the northern point of this meridian, at Sault St. Marie.

At Ft. Meigs, near Toledo, Ohio, the three men boarded a small schooner. Only three weeks earlier the Treaty of Ft. Meigs had been signed with the Indians granting concessions in Michigan Territory. Severe headwinds forced the men to complete their journey to Detroit on foot. On the outskirts of town they boarded at Williams Inn for $1.00 apiece. During the next three days they explored Detroit, but, unfortunately, William Austin did not record what he saw. Michigan Territory had impressed him, however, and soon after he arrived home he began making plans to return.

In 10 weeks Burt had covered over 1,000 miles andhad done little to ease his restlessness. He had seen for himself the beauty and richness of the Michigan Territory that would soon become home to a wave of immigrants attracted by the promise of new opportunities. He wanted to be a part of it.

Preparation

With renewed enthusiasm William Austin continued his partnership with John Allen, building mills whenever possible in Erie County. When the U.S. mail service expanded to the area, Burt became the first postmaster in Wales Center. Extra income was needed to feed his family which had grown larger with the birth of Austin on August 29, 1818, and Wells on October, 25, 1820.

His trip west had been a growth experience. It calmed any fears he might have had about exploring the unknown wilderness. It may also have strengthened his interest in surveying as a career. Such work demanded the scientific knowledge which he had been acquiring. Besides, there was the exhilaration of exploration, and, as the anticipated western migration developed, the work should be steady.

In 1818, the U.S. land surveys resumed in Michigan Territory, and the first land sales were made. A minimum purchase was 160 acres, one-quarter section, at $2.00 per acre. The price was payable one-fourth down and the balance in four annual payments, with no interest charged.[15] If the seller paid with cash in full, the per acre price was reduced to $1.64 per acre.

It was also in 1818 that travel to the West was greatly improved with the first steamship on the Great Lakes, *Walk in the Water,* offering passenger service between Buffalo and Detroit in just 44 hours.

With these incentives, William Austin made the decision in 1822, at age 30, to return to Michigan in hopes of landing a job in the public land surveys. Carrying letters of introduction from prominent Buffalo officials to Michigan's territorial governor Louis Cass and Congressional delegate William Woodbridge, Burt sailed from Buffalo to Detroit. He took the second steamship on the Great Lakes, *Superior,* which was powered by the engine from *Walk in the Water* that had broken apart in heavy seas the previous year.

His first stop was at Surveyor General Tiffin's office, in Chillicothe, Ohio. Unfortunately, there were no surveying jobs available. U.S. Deputy Surveyor Joseph Fletcher had already contracted with Tiffin to survey a 10,000 acre tract near Detroit, and his assistants would be selected from a long list of competent men that included John Mullett, Hervey Parke, and Lucius Lyon.

John Mullett had been recommended and placed on the list only three months earlier.[16] A tailor by trade, Mullett enjoyed mathematics more than textiles and taught himself the art of surveying.

Hervey Parke had been accepted only the previous year.[17] He, too, had been a restless New Yorker who had traveled to Michigan to find work in the land surveys. He was more fortunate than Burt, however, finding work as an assistant on a survey team. Afterwards, Parke was promised more permanent work. So, on the strength of this promise, Parke returned home, sold most of his furniture, and moved his family to an area north of Detroit.

Lucius Lyon, originally from Vermont, came to Michigan in 1821 and began teaching in a seminary school.[18] From a friend he learned about the U.S. land surveys, and his name was placed on the list of applicants. In 1823, Lucius Lyon was appointed a U.S. Deputy Surveyor.

Burt's narrowly missed opportunity probably postponed his career by more than a decade, but it did not dampen his determination to move to Michigan. When at Chillicothe, he caught the steamship Superior bound for Detroit, to find a suitable location for a future home. Aboard Superior Burt met two men who would become important contacts for him. One, a Mr. Webster, was anxious to construct a mill at Auburn, in Oakland County. The other was Hervey Parke; this was the first meeting of their eventual 36-year friendship.[19]

Thanks to Webster, Burt & Allen, Millwrights, was back in business. John Allen received William Austin's letter to come to Michigan immediately, and together they completed Webster's mill. Afterwards they contracted with Alpheus Williams to build a mill near Pontiac, but when that job was finished they were once again unemployed. William Austin then remembered that surveyor Joseph Fletcher was sub-dividing townships north of Pontiac, so he and Allen followed an Indian trail to a small trading post, now Flint.

There they were told that the surveyors had only recently passed through. Finally, two days later Burt and Allen reached Fletcher and his crew but found there were no jobs available. For awhile, Burt just watched as the crew methodically drove stakes into the ground at precise distances. So that was how it was done!

The earliest surveys were completed in the choice areas where settlement was most likely. Burt and Allen, therefore, headed south, towards Detroit, over land already subdivided. Because it would be several years before roads would be built, they followed Indian trails to an area north of Stony Creek village in Macomb County. There they selected land for their future homes.

The settlement, known as Mt. Vernon, was located in Washington township, about 26 miles north of Detroit. Land sales in the area had begun only the previous July. Together they purchased the southwest quarter of section 31, 160 acres, for $320. Burt took the northern half of the parcel. The land patent, signed by President James Monroe, was dated November 27, 1822. The purchase probably consumed most of their money, because they walked back home, through Canada in heavy snow, instead of taking the steamer.

For Burt, the bitter cold of winter was offset by the warmth of excitement in having taken a most important step toward a more promising future. During the next 15 months Burt and Allen continued their trade in Erie County, while making preparations for their eventual move to Michigan.

Chapter II

OPPORTUNITY IN MICHIGAN TERRITORY

Know ye the land of the billow and breeze,
That is pois'd, like an isle, amid fresh water seas;
Whose forests are ample, whose prairies are fine,
Whose soil is productive, whose climate benign?
Remote from extremes - neither torrid nor cold,
'Tis the land of the sickle, the plough, and the fold;
'Tis a region no eye e'er forgets or mistakes;
'Tis the land for improvement - the land of the lakes.

- from "Michigan," Detroit Gazette, August 6, 1824

Settlement

As the steamship Superior entered Detroit harbor on May 10, 1824, a wave of adrenalin surely surged through the veins of William Austin as he stood on deck with Phebe and their four children. A new chapter in their lives was beginning.

In Detroit, Burt loaded his family and their worldly possessions into a wagon provided by Hervey Parke. For the next several weeks they stayed with the Parkes until Burt could build a log cabin on his land in Macomb County.

His skill at constructing mills made his new home a structure that was sturdy and large enough to accommodate their growing family. It contained a fireplace, separate master bedroom, and a loft where the children and a hired hand could sleep. William Austin also built a blacksmith's shop with logs cut from trees on his property. Nearby, a large barn was added to shelter their farm animals. He built a forge to shape the various tools that would be needed.[20]

By now two of his boys had grown into useful workers. John, 10, and Alvin, 8, were old enough to swing an ax and help with the more difficult chores. A near-tragedy occurred in 1825 when Austin, age 6, nearly drowned in Stony Creek.[21] He became lost while returning from a neighbor's house and was forced to cross the stream on a log. He slipped into the icy water, which rose above his waist, but somehow he managed to wade across and safely reach his home.

Phebe cared for the two younger boys and made sure the family kept warm during the frigid winter months. After the flax was sown and harvested, Phebe spun it on her loom. Combined with the wool sheared from their sheep, it formed all the yarn they needed for the family's clothes.

William Austin continued his millwright trade to provide his family with the necessities of life. At age 32, he stood about five feet, 10 inches tall, about one inch taller than his father.[22] His strenuous work had contributed to his solid physique. During the summer of 1824 he built the Taylor and Millard mills on lower Stony Creek, and the following year he erected the Hershey mill on upper Stony Creek. The latter was in operation until about 1900. The Wadhams mill, located about six miles from Port Huron, was another example of Burt's craftsmanship.

4.   William Austin Burt *courtesy National Cyclopedia of American Biography*

Politics

When the Erie Canal was completed in 1825, land sales in Michigan Territory increased, although they lagged considerably behind those in Ohio, Indiana, and Illinois. On March 3, 1823, to help establish political order, Congress granted Michigan additional political rights which included the establishment of a Legislative Council. From a group of 18 men elected by the highest vote of the electors, the President of the United States appointed nine to form the Legislative Council, to enact legislation, and to make executive appointments. In 1825, the Council was increased to 13 members, chosen from an elected group of 26 people.[23]

William Austin was one of four persons elected by the people of Macomb County to be considered for the Council. From this group, he and John Stockton were appointed by President John Quincy Adams on March 9, 1826, to serve on Michigan Territory's Legislative Council.[24] The official term of the Council was two years. The two concurrent sessions of the Council, however, were active for only about six months. The first convened on November 2, 1826, and adjourned at the end of the year. The second convened on January 1st and adjourned April 17, 1827. During the latter period more than 344 pages of legislation were enacted, an amount greater than a normal four year period.[25] Included were:

- a bill establishing the minimum age for marriage with parental consent at 18 for males and 14 for females.

- a bill revising the criminal code to provide for death by hanging for those convicted of murder.

- a bill setting a fine of $500, or one year in prison, for anyone convicted of adultery.

Several bills provided for the planning and construction of roads within Michigan Territory.

William Austin maintained his desire to become a U.S. Deputy Surveyor. Less than two weeks before his Legislative Council term expired in 1827, his friend John Mullett forwarded Burt's application to Samuel Williams, Surveyor General Tiffin's Chief Clerk since 1822. Accompanying it was Mullett's personal letter of recommendation:

> I know of no one better qualified than Mr. Burt. He is honest and intelligent, well acquainted with surveying, both theory and practice. He . . . is universally respected."[26]

This was a profound vote of confidence because Tiffin considered Mullett his "principal deputy" whose "character for integrity and uprightness is unimpeached."[27] Mullett's efforts, however, were futile because the public surveys in Michigan were then nearly halted. This was not the time to be adding any more deputy surveyors.

Battle Creek

It was in 1825 or 1826 when John Mullett's survey crew became involved in a skirmish that later gave Battle Creek, Michigan, its name. One day while Mullett was subdividing townships near a creek in Michigan, two Indians entered his camp where the cook and a packer had remained behind. When the cook refused the Indians demands for food, a fight ensued. One Indian fired his rifle at the packer, but missed.

When Mullett and his crew returned they found one Indian tied up and the other unconscious. After freeing the Indians, Mullett felt it wise to break camp and head for home. A few weeks later the surveyors returned to finish the survey. They called the stream near the area of their encounter with the Indians "Battle Creek", thereby creating a name for the "breakfast food capital of the Nation."[28]

Masons

The Masonic Order is the oldest and largest secret society in the world. First introduced in America by English Masons about 1730, its membership included such notables as George Washington, John Hancock, and Paul Revere. William A. Burt joined the order in 1816, and in January 1828 he helped to form the first Masonic lodge in Macomb County, the third in Michigan Territory.[29] He was installed as the first Grand Master of the Stony Creek Masonic Lodge, which later became Rochester Lodge #5.

William Austin's leadership skills and courage were tested in a confrontation with Judge Samuel W. Dexter, founder of the town of Dexter in 1824.[30] As the town patriarch, Dexter was highly influential. In 1828, Dexter renounced his membership in the Masonic Order. He then became the editor of the *Western Emigrant,* an Anti-Mason newspaper. The Anti-Mason movement first gained strength in 1826 when a few of the dissident members conspired to publish some of the Mason's closely guarded secrets. When one of the Anti-Masons, William Morgan, disappeared, the Masons were blamed.

During this period of unrest, in 1828, William Austin contracted to build a mill in Dexter, about nine miles northwest of Ann Arbor. He was busy at work when several Anti-Masons presented him with a challenge to debate Samuel Dexter on the Masonic issue that evening in the local schoolhouse. Burt accepted, and, according to at least one source, "Burt came off victorious, much to the chagrin of Dexter."[31] Burt's coup de grace cited Dexter's withdrawal from the Masonic Order. He appealed to the audience to determine, if they could, at what time it was when Dexter was an honest man:

> Was it when he voluntarily took an obligation to abide by and sustain the Order, or was it when he violated that obligation and denounced it.[32]

With the last board in place on the Dexter mill, William Austin uncorked a barrel of whiskey he had hidden in his wagon. It was then customary to toast the completion of important structures, and on this occasion there was the additional satisfaction of his successful debate with Dexter. Although he practiced temperance, William Austin also believed in tradition.

The Anti-Mason party became the first third-party challenge to the traditional two-party system in American politics. In 1831, Samuel Dexter bid to represent Michigan Territory in Congress on the Anti-Mason ticket but was defeated by Austin E. Wing, an active Mason. With the support of Henry Clay, the Anti-Mason party entered the election of 1832; however, President Andrew Jackson, a Mason, was re-elected by a landslide. Eventually the Anti-Masons were absorbed into the new Whig party.

Chapter III

AMERICA'S FIRST TYPEWRITER

Necessity, who is the mother of invention.[33]

- Plato (428 - 347 B.C.)

Burt's Writing Machine

Following an exhausting tour of duty on Michigan's Legislative Council in 1827, William Austin returned home to find an imposing volume of mail that had accumulated during his absence. His reputation had grown considerably since his election in 1825, and his advice was frequently sought. There were, however, other priorities to attend to. For one, his fifth son, William, was born October 31, 1825, and his family obligations required considerable attention. His correspondence would have to wait until a faster method of writing could be developed. The old quill pen and ink method was just too slow.

Nearly two decades earlier, in 1809, William Austin had devised a system of shorthand to aid in more rapid note-taking during his personal study.[34] Letter writing, however, could not be simplified. The problem of keeping up with his correspondence mounted until Burt finally conceived a mechanical solution in 1828.

Thus, unlike many who invent for wealth, William Austin Burt invented America's first typewriter out of a genuine need to resolve a problem. Once he formulated the concept, he began work in earnest on a functional model. Many of the parts and tools had to be forged in his own workshop. The type was supplied by his friend John Sheldon, editor of the *Detroit Gazette*, a precursor of the *Detroit Free Press*. The *Gazette* was one of Michigan's most influential newspapers, and Sheldon's articles frequently sparked political controversy. Sheldon, a co-founder of the newspaper in 1817, also served as clerk of Michigan Territory's first three Legislative Councils. It was probably during Burt's term on the Council that the two first met.

While Burt was building his writing machine, Sheldon became embroiled in a classic confrontation involving the freedom of the press.[35] In January 1829 Sheldon criticized Michigan's Territorial Supreme Court for its decision in a larceny case. When the court demanded he remain silent, he responded with scathing editorials and was promptly fined $100 dollars. Instead of paying the fine he chose a nine-day incarceration in the Detroit jail. By then his editorials had so aroused the citizens of Detroit that 300 of the 2,200 total residents of the city staged a massive dinner in the jail. Collectively they displayed their support for Sheldon by raising bail for his release.

Burt, meanwhile, completed a crude and simple machine, which actually worked.[36] The mechanism was housed in a small wooden box measuring 12 inches wide, 12 inches high, and 18 inches long. On the front was a circular clock-like guage to indicate the distance traveled by the paper as it moved across the impression bar. A single sheet, up to 15 inches long, was fastened to a belt covered with a velvet-like material. The belt carrying the paper was rotated each time the impression lever was depressed.[37]

Sheldon's type was fastened to a semicircular segment of the underside of a pivoting arm. The type characters included both upper and lower case letters, as well as the hyphen, comma, semicolon, dash, and period. Each character was listed on a curved scale placed along the top front of the box. To operate the machine, the desired character was selected by inserting the tip of the pivoting arm into a groove opposite each character on the scale. When a lever was depressed the inked type, filled by special pads, contacted the paper.

With practice Burt could type out a letter that looked surprisingly neat. The machine, however, was painfully slow to operate, and Burt was falling further behind in his correspondence. Despite this flaw, John Sheldon was ecstatic when he saw Burt's writing machine. He envisioned one in every home and business. With a little luck and successful promotion, both Burt and Sheldon would be rich. William Austin was less optimistic, but he was encouraged by Sheldon's enthusiasm. After all, he had a large and growing family to support.

On April 23, 1829, Sheldon took a leave of absence from the *Gazette* for "health reasons."[38] His partner, Ebenezer Reed, took over as temporary editor. During the next several months Sheldon worked closely with Burt to promote the new writing machine. Burt had hesitated to select a name for his invention. He called it a Typographer, but Sheldon suggested he create widespread interest by seeking names from readers of several newspapers. In May, an editorial appeared in the *New York Commercial Advertiser* describing Burt's invention as "a simple, cheap, and pretty machine for printing letters. It should be found to fully answer the description of it." The best suggestion received was "Burt's Family Letter Press," but William Austin preferred to stick with "Typographer."[39]

To Burt's knowledge a writing machine had not previously been conceived or patented. As he prepared his patent application in May 1829, Burt wrote:

> To All to whom these Presents shall come:
>
> Be it known that I, William, A. Burt, of the township of Washington, in the county of Macomb and Territory of Michigan, have invented a new and useful machine styled by me "the typographer", to be used by individuals, for printing with but one type of a kind.

His specifications included a proposal for a "compound Typographer" that could make "several copies of the same writing at one operation."[40]

Sheldon typed out the following letter of application to the Secretary of State, Martin Van Buren:

> Detroit, M.T.
> May 25, 1829
>
> Sir: This is a specimen of the printing done by me on Mr. Burt's Typographer. You will observe some inaccuracies in the situation of the letters: these are owing to the imperfections of the machine, it having been made in the woods of Michigan, where no proper tools could be obtained by the inventor, who, in the construction of it, merely wished to test the principles of it, therefore, taking little pains in making it. I am satisfied from my knowledge of the printing business, as well as from the operation of the rough machine, with which I am now printing, that the typographer will be ranked with the most novel, useful and pleasing inventions of the age.
>
> Very respectfully,
> Your obedient servant,
> JOHN P. SHELDON

On July 23, 1829, President Andrew Jackson and Secretary of State Martin Van Buren signed the Letters Patent for Burt's Typographer. It granted William Austin Burt, for a period of 14 years, "the full and exclusive right and liberty of making, constructing, using, and vending to others" America's first writing machine. During the following months Burt's enthusiasm for his invention waned, while Sheldon remained optimistic.

5.   John P. Sheldon (1792 - 1871)
      courtesy of The State Historical Society of Wisconsin

6.   Burt's Typographer - America's first typewriter.
courtesy of the Smithsonian Institution

Sheldon contacted a Mr. White at a New York foundry for information regarding costs of various type for use in the Typographer. He was advised that

> . . . the cheapest way for your friend Mr. Burt to get on would be to come to this place and direct in person what he wants done. He here would find tools and materials for bringing his machine to perfection, and if it prove a good thing when done he will be able to dispose of the most here. Should he come here I will assist him in getting on, if requisite.[41]

Burt agreed with the recommendation and, on October 29, 1829, Sheldon wrote to his wife:

> Mr. Burt informs me that he will leave this for New York, on Wednesday next . . . Burt's prospects, in relation to his invention are good; and he has offered me one half of the proceeds for my assistance in getting it into vogue. There have been a considerable number already spoken for in this territory—but it is Burt's intention to dispose of the rights for states, counties, etc. and not establish a manufactory.[42]

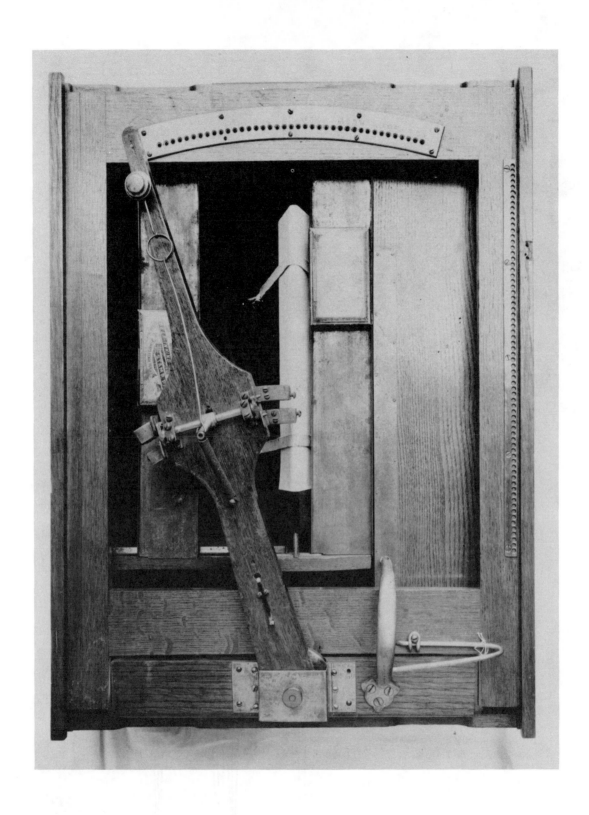

7.   Burt's Typographer - Top view
courtesy Smithsonian Institution

8. Demonstration of
William A. Burt's typographer
courtesy Compton's, division of
Enclopedia Britannica,Inc.

W. A. BURT.
TYPOGRAPHER.

Patented July 23, 1829.

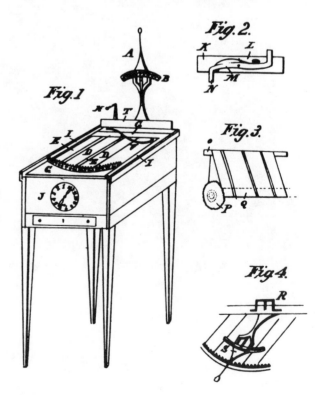

9. Restored patent office drawing
courtesy National Archives

10. Letters Patent - July 23, 1829
courtesy National Archives

Nearly five months elapsed before an improved model was built. It has been described as "beautiful, . . . the size and shape of a modest pinball-game machine with four gracious, tapered, Hepplewhite-style legs, and including several mechanical improvements."[43] Unfortunately, the machine was as slow to operate as the original. Short of funds and homesick, Burt typed the following letter to Phebe:

New- York March, 13, 1830

Dear Companion,

I have but jest got my second machine into opperation and this is the first specimen I send you except a few lines I printed to regulate the machine , I am in good health but am in fear these lines will not find you so andthe children from the malencholley account your letter gave me of sickness and deaths in our neighbourhood,, I had rested contented to what I should if it had been summer seasen adout the health of my family, as it is jenerlly healthy during the winter months; but

their has ben an unusual quentity of sickness heare this winter, and it has ben verry cold in Urope as well as in America, a strong indication of the change of seasonth that I have so often mentioned.— Mr Sheldon arrived here four days ago he went imediately on to Washington and took my moddle for the Pattent Office, he will returne here next week at which time I shall put my machine on sale and shall sell out the patent as soon as I can and return home, at aney rate I seall returne home as soone as the Lake nevigation is open if life and health is spared me. I have got along but slow since I have been here for the want of cash to hire such help as I wanted; I have been as prudent as I could, have taken my board with a family from MCyuga who keep a boarding house they are verry good christian people and are kind to me. I pay three Doll ars a week for my board.—You must excuse mistakes , the above is printed among a croud of people asking me many questions about the machine. Tell the boys that I have some presents for them. If I had aney news to communicate I would print more but as I have none I must close hopeing these lines will find you well.I wish you to write as soon you receive this, do not make aney excuses I shall like see it in aney shape

William A. Burt.

Phebe Burt

11. Burt's letter to his wife Phebe
courtesy Smithsonian Institution

20

Sheldon had taken the original crude model of the typographer to the Patent Office. It was later destroyed in the Patent Office fire of 1836, but a reproduction was built by Burt's great grandson Austin Burt in 1892 for the Columbia Exposition in Chicago.[44] The ultimate fate of the second writing machine is unknown. The typewriter, however, had attracted the interest of at least one investor, Cyrus Spalding. He gave William Austin $75 for the right to manufacture the Typographer. On March 17, 1830 Burt typed a letter to Spalding with suggested improvements for constructing the machine:[45]

> I have by way of experiment, got some type cast on bodies of a size suitable to set one after an other in a circular groove as was mentioned to you while at this place; the alphabet can be made in this way to answer the purpos, and should be made of harder mettel then type are usually made of an addition of copper and zinck will make them harder. . . . Please inform me how you get along.- Do not take this as a fair specimen of printing, from my new machine. I have had a boy to work on it a day or two past he has got some what out of order.

Apparently Spalding became disillusioned. Eight years later, in a letter dated June 10, 1838, he informed Burt:

> Sir, after spending considerable time and money trying to build a suitable machine and one that would be useful, I gave up. After asking the advice of some of our best mechanics we came to the conclusion that it never would be useful and it would be very liable to get out of order which was another objection, and it would require a mechanic to keep it in order so as to make good work.[46]

Because he had not actually made any of the machines, Spalding thought it would be "right and just" if he could have his $75 back. It is unknown whether Burt ever repaid Spalding or acknowledged his letter, but it is probable that Spalding eventually learned what Burt already knew: There was no market for writing machines in the early 19th century. The U.S. mail service was developing, but it was inconsistent and slow. Even if there had been a widespread demand, there were no adequate manufacturing facilities available to produce the machines economically.

The historically unique aspects of William Austin Burt's Typographer are intact. It was the first typewriter that printed with both upper and lower case letters, a feature not included even on the early mass-produced typewriters at the end of the nineteenth century.

Although Burt was no doubt unaware of them, other writing devices had been conceived in Europe prior to 1800. On January 7, 1714, Queen Anne issued a patent to Henry Mill, an English engineer, who claimed that he had:

> . . . by his great study, paines, and expense, lately invented and brought to perfection an artificial machine or method for the impressing or transcribing of letters singly or progressively one after another, as in writing, whereby all writings whatsoever may be engrossed on paper or parchment so neat and exact as not to be distinguished from print. . .the impression being deeper and more lasting than any other writing, and not to be erased or counterfeited without manifest discovery.[47]

Mill's device may not have been a machine at all, but rather an embossing device intended for use by the blind. No drawings or specifications have ever been found, and it is doubtful whether an actual machine ever existed.[48] One British author, however, contends that Mill built one model for demonstrations, and that it had an inked ribbon.[49]

Unquestionably, William Austin Burt invented America's first typewriter, but who invented the world's first typewriter? That question may never be answered. Each country claims to have a candidate: Von Knass of Vienna in 1753, Jaquet-Droz of Switzerland in 1772, Pingeron of France in 1780, and Turri of Italy in 1808.[50]

William Austin Burt could scarcely have envisioned the evolution of writing machines that would follow the Typographer, and at the time he was not particularly interested. Within three months after he had patented the machine, he decided to sell his rights. Since the Typographer did not resolve his correspondence problem, he moved on to other things. He had spent nearly a year working to perfect and promote his invention. With Michigan in the midst of an economic depression it was time for William Austin to start earning a full-time living. Keeping a wife and five growing boys well fed was not an easy task. Inevitably, Burt turned his attention again to surveying.

Chapter IV

A NEW CAREER

*And the city is laid out as a square, and
its length is as great as the width; and
he measures the city with the rod, fifteen
hundred miles, its length and width and
height are equal.*

- Revelation 21: 16

Preparation

After his prolonged stay in New York during 1830, William Austin returned to Mt. Vernon. There were no offers for surveying jobs, despite Mullett's earlier recommendation. At age 38, Burt had yet to obtain a single U.S. government surveying job. If he was going to be a surveyor he would have to start at the local level. In 1831, he was elected surveyor for Macomb County. The following year his duties were expanded with his appointment as district surveyor by Michigan's territorial governor. At last, he was gaining his overdue experience.

On December 19, 1832, Burt was appointed Mt. Vernon's first postmaster.[51] Once a week his son Austin rode into the village on horseback to collect the mail for the neighborhood.[52] In those days, however, there were few letters written as few could afford the postage of 25 cents.

On April 23rd William Austin became an associate justice of the Macomb County Circuit Court, and soon acquired his lifelong title of "Judge Burt."Between jobs William Austin taught surveying to his five sons, four of them active teenagers ranging in age from 13 to 19. He also accepted construction work whenever possible. The last mill William Austin built was at Frederick, near Mt. Clemens.

In July 1833 Burt's friend Lucius Lyon was elected a delegate to Congress by Michigan's territorial legislature.[53] The following month Lyon informed Surveyor General Micajah Williams of a letter he had received from William Austin with regard to the land surveys in Michigan north of Macomb and St. Clair counties. Burt had written:

> I should be pleased to obtain a district to survey there. I am informed by hunters who have explored that section of country that a large portion of it is first rate land.[54]

Lyon recommended Burt to Williams as "a gentleman of high standing in the county where he resides and his statements may in all cases, be confidently relied on."[55]

At last, William Austin's determination and Lucius Lyon's influence were successful. On November 23, 1833, Burt was appointed a U.S. deputy surveyor. At age 41, William Austin Burt was about to begin his real career.

12.   Lucius Lyon (1800 - 1851) *courtesy, Michigan State Archives*
(from MICHIGAN HISTORY MAGAZINE, 1923, VII, 19)

U.S. Public Land Surveys

With the signing of the Declaration of Independence in 1776, the colonies ceded their lands to the U.S. government. Proceeds from the sale of these lands were then used to pay off Revolutionary War debts and to reward soldiers who fought in that war.

Early settlers in most of the original 13 colonies, as well as in Tennessee and Kentucky, followed the indiscriminate "metes and bounds" method of locating boundaries. Shortly after the Revolutionary War ended, a group headed by Thomas Jefferson proposed a more orderly system of rectangular surveying, used centuries earlier by the Romans, who held that land division by meridians and parallels was based on a religious and mystical foundation.[56] The Land Ordinance of 1785, approved by the Continental Congress on May 20th, established the legal basis for our public land surveys over the past 200 years. It contained several important provisions:[57]

- Land must be purchased from the Indian inhabitants before it can be surveyed. This policy had been followed in the colonies.

- The land must be surveyed before it can be sold to settlers by auction. Several statesmen, including Alexander Hamilton, favored the old metes and bounds method that did not include this requirement.

- A qualified surveyor will be appointed from each state to divide the territory into townships six miles square. The north-south lines are to be run with east-west lines crossing them at right angles. Several meridians (north-south lines) will be established, each with an intersecting baseline (east-west line). The location of all land will be referenced from its' particular meridian and baseline.

- Each township, measuring six miles square, is to be marked by subdivision into sections.

- Sections, originally called lots, are to be one mile square (640 acres). They will be numbered from one to thirty-six.

- The lines shall be measured with a chain, plainly marked by chaps on the trees, and exactly described on a plat (a surveyor's map, which may show topographical and other features of the land).

- The geographer and surveyors shall pay the utmost attention to the variation of the magnetic needle. They shall run and note all lines by the true meridian, certifying with every plat the variation noted at the time of running the lines.

TOWNSHIP GRID

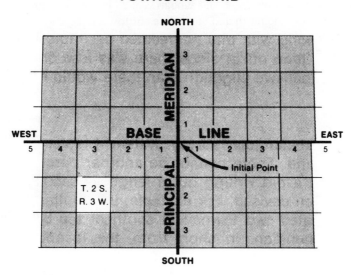

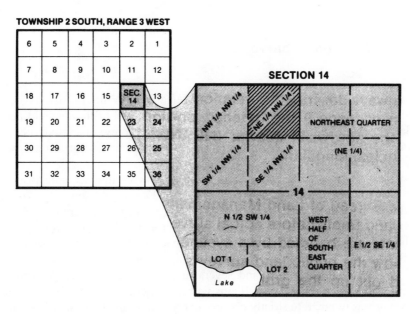

13. The Rectangular System of Land Division
courtesy Bureau of Land Management

25

The Ordinance of 1785 led to many surveying problems. Most were worked out with subsequent revisions. One significant flaw was that no provision was made for the spherical shape of the earth which causes north-south lines to converge near the poles. As a result, a tract of land is really never square, nor do lines intersect at perfect right angles. Later, correction lines were established every 24 miles to compensate for this convergence, with a greater offset as the lines were extended poleward. An offset near the equator may be only a few feet, while one in northern Alaska might be a tenth of a mile.

Another problem was the requirement that lines be run by the true meridian. Because of delays in the first surveys, Congress waived this requirement in 1789 and permitted use of the magnetic meridian. This resulted in many lines varying from their intended positions.[58] On May, 18, 1798, by Act of Congress, surveyors were again required to run the lines of the public lands by the true meridian. Rufus Putnam, Superintendent of Surveys for the Ohio Company, complained to Congress, in 1798, that:

> . . . it will be extremely inconvenient and embarrassing, if not altogether impracticable for the deputy surveyors to run the north and south lines of townships, etc., according to the true meridian . . . there is a difference in the variation of the magnetic needle from the true pole, in different places, and those at no great distance from each other . . . so that a compass rectified or adjusted to the true meridian in one place will not cut that meridian in all parts of the territory . . . it will be necessary to take frequent observations.[59]

In 1833—the year William Austin Burt was appointed a U.S. Deputy Surveyor—the General Instructions of the Surveyor General "for the states of Ohio and Indiana, and the Territory of Michigan" included the provision:

> All lines, of whatever description, which you may survey, must be run by the true meridian.

As Burt would soon find out, the problem of running accurate lines remained an inconvenience and embarrassment to the surveyors.

W.A. Burt's First Government Survey

William Austin's first contract from Surveyor General Micajah T. Williams, dated November 25, 1833, was accompanied by a pocket-sized booklet that Burt would eventually fill with field notes from his first survey. It included Williams' instructions to "faithfully lay out, survey, and subdivide into sections" 12 townships north of Port Huron, bordering Lake Huron.[61] For this work Burt would be paid $2.75 for each mile of line run.

He had one month to form a party and begin his survey. Knowing that accurate work was expected of him, he carefully selected his crew of six men, including his 17 year old son Alvin. He would also take his time and do the job right, a mistake usually avoided by profit-minded deputies who had learned that the key to financial success under the contract system of surveying was to perform the greatest amount of work in the shortest possible time. It was also a principal cause of inaccurate surveys.

In December 1833, Burt packed into his wagon the tools of the trade, including a magnetic compass, surveying chain, tally-pins, marking iron, and several axes.[62] He also carried a copy of the "General Instructions to Deputy Surveyors" sent by Williams with the suggestion that it be committed to memory. Every deputy surveyor was required to strictly observe every detail, great and small. Plenty of food staples were added, such as pork and beans, which would simmer all night on the smoldering coals of the campfire—dried fruit, sugar, coffee, and flour. For the first few meals, however, they would enjoy some prepared food Phebe had sent along, a welcome treat that would be repeated on future survey trips.

After reaching their assigned district the surveyors went right to work. After positioning their open-sight compass on a tripod at the starting point, appropriate adjustments were made. The variation from true north was noted in the field notes. One of the men moved forward about 200 feet in the direction the line was to be run. He then set up a rod, a wooden pole that was aligned by the compassman with the sight vanes on his magnetic compass. The axeman cleared away any brush obstructing the line of sight, and the point was marked. The surveyors then moved forward to the mark, and repeated the process.

The chainmen followed behind, measuring the line with a 66 foot chain that contained 100 links, each precisely 7.92 inches in length. One historian described the role of the chainmen after the initial tally pin was placed at the starting point:

> One stood there, holding the chain loosely in five-foot loops, while the other carried the other end of the chain forward through the clearing hacked by the sweating axemen. At every half-chain length he set a tally pin in the earth, tagged with red cloth. When he encountered trees too big to waste time falling, he had to run a deflection around it, guided by Burt. As he approached the end of the chain, the rear chainman cried "Chain!" and the forward man went ahead cautiously, setting another tally pin at the end of the chain.[63]

After a length of 10 chains had been run, the lead chainman called, "Tally One". The rear chainman then came forward, counting the tally pins in the presence of the other, and took the forward end of the chain to begin another tally. By alternating positions the chainmen continued up the line.[64]

The axemen, under Burt's supervision, then marked the larger trees along the line with two notches. At each section corner a post was driven into the ground. Burt then selected two nearby trees and marked them. With a scribe he carved the town, range, and section, and the letters B. T. to identify them as "bearing trees". Their species, size, and exact location were recorded by Burt in his field notes.

Hervey Parke was running the exterior township lines in the area, and Burt was instructed to check his surveying chain with the "standard" carried by Parke. William Austin had been in the field less than three weeks, however, when Parke was forced to return home.[65] Parke informed the Surveyor General that his party had been able to survey only 100 miles in 40 days. The severe cold had formed a crust on the snow that could bear the weight of neither man nor horse.

Burt and his survey party were no more fortunate than Parke. In February 1834, with only two-thirds of his contract completed, William Austin was forced by flies, mosquitoes, and impassable swamps to abandon the work until later in the year. He had not prepared for these obstacles, and, although he did not complain about it, his first contract resulted in a financial loss. He would chalk it up to inexperience and would learn from his mistakes. Someday, he thought, he would write a practical manual to spare the novice surveyor such agony. The lack of basic information occasionally caused a new deputy to take shortcuts and, perhaps, even to falsify his field notes. For Burt, however, accuracy was an obsession.

Chapter V

QUEST FOR ACCURACY

While the genius of Franklin snatched the lightning
from the clouds to minister to man's insatiate thirst
for science, it remained for a Burt to seize the sun
beams and compel them to point out the true poles
of our earth.[66]

- John Wilson, Commissioner, GLO

Burt's Second Government Survey

Although Burt had not completed his first contract, Surveyor General Williams was impressed with his work. In a letter to Lucius Lyon he wrote,

> Your friend, Mr. Burt, proves to be an excellent surveyor. For a first contract he has returned the most satisfactory work I have yet met with.[67]

In September William Austin returned to an area northwest of Lexington to finish his work on the four remaining townships.[68] This time he was joined by two of his sons, Alvin and Austin, with Austin serving as cook for the party of five men. Blessed by fine weather the surveying job was completed within eight weeks.

Earlier, in the spring, William Austin surveyed the railroad line from Detroit to Ypsilanti. It was one of the first surveys for that purpose in Michigan Territory.[69] This was a period of railroad development for Michigan, and the Detroit & St. Joseph Railroad had been chartered two years earlier. In the fall of 1834, Burt and one son, probably John, began a survey for the railroad line from Detroit to the mouth of the St. Joseph River.[70]

Before the work was completed William Austin obtained his second contract from the General Land Office. He and 14 other surveyors, including Hervey Parke and John Mullett, were instructed to subdivide 212 townships in Wisconsin Territory. Burt's district of 13 townships extended from the present site of Milwaukee northward to the western shore of Lake Michigan.[71]

William A. Burt's survey party of seven men, including Alvin as axeman and Austin as chainman, left Mt. Vernon on December 1, 1834. With supplies and instruments loaded into a small horse-drawn wagon, the surveyors headed south, on foot, to Detroit. From there they followed the stage coach route to Chicago, where Burt acquired additional supplies and a four-horse team to carry them to Milwaukee. Chicago was then only a small community with a few houses along the mouth of the Chicago River. On Christmas day 1834 they arrived at Solomon Juneau's trading post on the east bank of the Milwaukee River. At last the land developers, who had urged that this area be quickly surveyed and opened to public sale, would be pleased.[72]

It is a tribute to William Austin Burt's abilities that he was selected to subdivide the township in which the heart of modern Milwaukee developed. Surveyor Hiram Burnham complained to the Surveyor General that "Judge Burt is fortunate in having good places for Deposit & Settlements on a good portion of his district."[73] Burnham's frustration was understandable, for he had averaged only eight miles a day, having to cut through dense brush and timber in snow 14 inches deep. Hervey Parke had even more difficulty, averaging less than four miles a day. Parke complained his assignment was "the most difficult and arduous district ever given out in Michigan."[74]

Hiram Burnham's problems had just begun. His surveying lines would not close, a phenomenon that had plagued nearly all government surveyors at one time or another. In 1826, there were complaints that several of the early Michigan surveys were erroneous.[75] Careless or dishonest surveyors were initially considered the cause, but, in most cases, the inaccurate work was caused by strange aberrations in the surveyor's compass needle. Burnham had encountered the problem the previous season, and apologized to the Surveyor General:

> I am fearful that you will not find the closing fully satisfactory; they are not so to me. I cannot account for the variations that occasionally occur, other than that local attraction is the cause. I have repeatedly doubted my compass & chainmen—have retraced my lines & caused them remeasured—but always with the same result. . . . I believe Mr. Park mentioned the circumstance of his having been troubled—All the streams passing through this region are more or less impregnated with Iron—And also in the swamps, on standing water an Iron coating is to be seen. But in no instance have discovered either bog or mountain ore—When found— I am of the impression it will be North & East my District—North of Mr. Burts—Extending South westerly—.[76]

Even Burt's friends Lucius Lyon and John Mullett had encountered the problem. In a letter to Surveyor General Williams in 1832, Lyon wrote:

> The variation of the compass here is so widely different at different places that very little dependence can be placed on the needle. More than half our lines have to be run a second time, so that, at best, our progress is slow.[77]

On another occasion, Mullett informed Williams:

> I find it very difficult to close the work with a tolerable degree of accuracy on account of the great difference of the variation of the needle. I have been obliged to go over most of my work, some for a second time and some more.[78]

It was reported that in 1833 Abraham Lincoln, while a deputy surveyor for Sangamon County, Illinois, learned that the compass needle was often deflected by daily variation, magnetic storms, or whenever an axe, chain, or hunting knife came too close to the compass.[79]

The Surveyor General's Instructions, issued to Burt and all other U.S. deputy surveyors in Ohio, Indiana, and Michigan Territory, warned of the problem:

> The aberrations of the needle, are a fruitful source of error in surveying. These may arise from a variety of causes. "Local attraction", owing to the presence of iron mineral, is generally assigned by surveyors as the principal cause of the disturbance of the needle.

So that this would not form the basis for future excuses, a caution was added:

> But it is believed that in many instances, the true source of the errors complained of, is to be found in the carelessness or inattention of the surveyor, in the use and management of his compass, or the erroneous measurement of his lines.[80]

14.   Abraham Lincoln, surveyor   *courtesy, GPO (Lola Gazier, SURVEY AND SURVEYORS OF THE PUBLIC DOMAIN), p.70*

In effect, surveyors were told that the problem was one to cope with rather than overcome. Now, however, in April 1835, Hiram Burnham was having difficulty even coping:

> The Abbirations of the needle in this Town Exceed any thing I have ever met with. . . In fact as I approach the Lake there is no calculating for it.[81]

One month earlier Burnham had found a piece of iron ore on the line and had it sent to Burt, who was working in an adjacent district to the north. Burt, himself, had experienced difficulty in closing his lines, but he did not inform Williams about it until May 17, 1835. Although Williams had formally resigned as Surveyor General, Burt wrote:

> I have yet to learn the name of your successor at aney rate you will know what to do with these Lines. . . . the aberations of the Neddle here is truly perplexing. verry maney of my N & S Lines have to be corrected.[82]

He sent a similar message to his wife Phebe, but added:

> It is most annoying this inability, as yet, to discover a method for doing away with the difficulty or the cause thereof.[83]

Indeed, when Burt attempted to close his lines he found that some of the intersections varied by as much as a full chain, or 66 feet, from the corner post.

When his contract was completed on June 5, 1835, Burt personally carried his field notes to Cincinnati to present them to the new Surveyor General, Robert Lytle. He then returned to Mt. Vernon to test an idea he had been formulating which just might overcome the problem of the unreliable magnetic compass.

Variation of the Compass

The Surveyor General's Instructions required that the lines be run by the true meridian, and the variation of the magnetic needle must be taken and recorded in the surveyor's field notes. The magnetic variation is defined as "the angle at which the compass needle points away from the true north."[84]

The "line of no variation", or agonic line, is aligned in the meridian of the magnetic needle and passes through the north and south magnetic poles. Along this line the magnetic needle coincides with the true meridian. Moving westward or eastward from this line causes the needle to decline from the true meridian, and this "declination" is often called the "variation" of the compass. This variation is irregular, and the line, itself, continually moves east or west from time to time. With the variation of the compass at any place continually changing, it was most important that the surveyor accurately record the variation at the time of the survey.

Surveyors were instructed to determine the variation of the magnetic needle by observing the north star:

> . . . note exactly . . . when the Polar Star and the first in the tail of The Great Bear are vertical to each other, or range with a Plumb line. Ten minutes afterwards the Polar Star will be on the true meridian a Stake or Small light placed at that time in a right line with the Plumb line and the Polar Star will point out a true meridian, To which the Compass being applied will s(how) the variation of the needle.[85]

The observations could be made only at night, and the process was extremely slow. In practice, the results were often inaccurate. Section lines run one mile apart could have different variations, due to a change in local attraction. Unless a better method could be devised to make more frequent determinations of the variation of the compass, accurate surveys would be impossible.

The Solution

The frustrations William A. Burt encountered in Wisconsin convinced him that the magnetic compass could no longer be relied upon for surveying the public lands. Other surveyors had reached that same conclusion, but it was Burt who first developed a reliable solution. He, more than the others, possessed a unique blend of mechanical genius, a quest for accuracy, and a Yankee determination to solve problems.

It was Burt's knowledge of astronomy that enabled him to formulate the idea for a compass that did not depend upon the magnetic needle for direction. He understood the ancient concept that had guided generations of explorers on both land and sea: There is a regular relationship between the sun and the earth. The sundial, possibly one of the earliest of scientific devices, operates on this same basic principle to determine the time of the day.[86]

Burt recognized that a surveying instrument could be designed to use the position of the sun to locate true north from particular spots on the earth. Such a device required three essential parts to accurately determine the meridian:

- A "declination arc", to set the location of the sun north or south of the equator at the particular time and date of the observation. The value, expressed in degrees, is obtained in an "ephemeris" (or nautical almanac), a publication issued periodically that includes a table of the sun's position at a specified Greenwich mean time and date.[87]

- A "latitude arc," to set the location, or latitude, of the observer.[88] The latitude, also expressed in degrees, is the distance north or south of the equator, and is determined by taking a noon observation of the sun.

- An "hour arc," to set the time of the day.

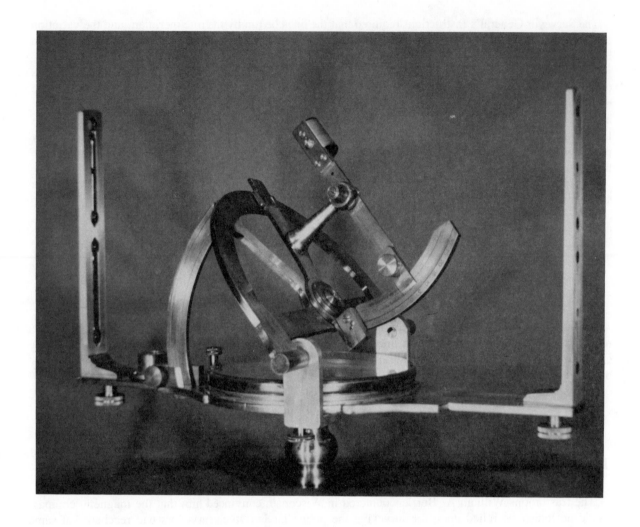

15. Reproduction of Burt's original variation compass - patented 1836
courtesy, Smithsonian Institution.

In July 1835 William Austin Burt began constructing a model for a new compass that incorporated these principles of astronomy. As soon as all the parts were assembled and properly adjusted, his device was attached to a common magnetic compass to permit the following determinations to be made:

- The true meridian (north-south line through the true poles).

- The variation of the compass (the difference between the true meridian and magnetic north.)

- The time of day, as shown on the scale of the hour arc.

- The latitude of the instrument's position.

Although a layman may struggle with the concepts used in Burt's compass, a contemporary surveyor has written that "finding north with William Burt's device is almost as simple as lighting a match with a magnifying glass."[89]

On August 10, 1835, Burt notified the Surveyor General at Cincinnati that he had successfully completed some experiments for a unique compass, details of which would be forwarded to him in the near future.[90] William Austin then completed a crude model of his invention and sent it to William J. Young, a Philadelphia instrument maker, with instructions to make a workable apparatus.

Born in Scotland in 1800, Young served his apprenticeship from 1813 to 1820 with Thomas Whitney, a prominent Philadelphia instrument maker. In 1826 he established his own business and soon became the first American instrument maker to develop a capacity for mechanical graduation of instruments. In 1830 he invented and commercially produced the surveyor's transit.[91] Undoubtedly it was Young's reputation for high quality precision work that caused Burt to select him to build his first compass.

Young was also a member of the prestigious Committee on Science and the Arts of the Franklin Institute, considered the "most important technical organization in the country."[92] Young submitted Burt's new compass for their scientific review, but to avoid any possible bias he did not participate in the committee's evaluation. The final report, adopted December 10, 1835, was completed by Alexander Dallas Bache, a great-grandson of Benjamin Franklin and a noted American educator and physicist.[93] The committee defined the invention as:

> an instrument to ascertain the variation of the compass, invented by Wm. A. Burt of Michigan.

Bache then described how to use the compass:

> When the time of day & the latitude of a place & the sun's declination are known nothing more is necessary to find the variation than to set the equatorial circle to the elevation of the equator, the hour arm for the time of day, the declination arm for the sun's declination at the given time, then turn the whole compass in a horizontal plane until the image of the sun given by the lens falls between the parallel lines before referred to. The line of sight will then be in the meridian and the variation will be read off by the needle.

Although the apparatus could be removed from the common compass by loosening clamp screws, the committee suggested it would be best if the two units were permanently attached.

The reviewers concluded their report:

> The committee believe the instrument of Mr. Burt to be new. They consider it highly ingenious, susceptible of considerable accuracy in its use, and as giving results with great facility. They therefore recommend to the Managers of the Franklin Institute to award Mr. Burt a Scott's Legacy Medal and a premium of twenty dollars.[94]

Even before the Franklin Institute reviewed his invention, William Austin was making plans to have it patented. In November he traveled to Washington and located a draftsman near the U.S. Patent Office who would attend to the details. William Young sent the model that would accompany the patent application, for which he charged Burt $25. It was of the same size and pattern as Burt's original model.[95] In early February 1836, William Austin acted to avoid any possible delay in the patent approval. He urged that the draftsman immediately contact his friend Lucius Lyon if any additional charges were needed, and Burt would reimburse Lyon later.

Lyon was in Washington, D.C. to represent Michigan as one of its first U.S. Senators. Michigan's statehood was delayed, however, because of a boundary dispute with Ohio so he could not vote in the session that began December 1835. Although Lucius Lyon could only participate as an observer until January 26, 1837, he was not without influence. He was, in fact, one of the only politicians who recognized the potential of the Upper Peninsula.

On February 26, 1836, William Austin Burt was issued Patent No. 9428X for his invention "for determining the variation of the needle, the true meridian, and the apparent time." By then the device had successfully passed its first test. Young completed the first "variation apparatus" in November 1835, and the instrument was tested by Burt's son Alvin in the spring of 1836, while he was subdividing townships west of Milwaukee.[96] Its value in taking frequent and accurate variations was confirmed. In June 1836 William Austin had an opportunity to test the compass while continuing the surveys near Milwaukee.

Young sent William Austin the first "variation compass," on July 13, 1836. It was an improvement over the original "variation apparatus." As suggested earlier by Bache's committee of the Franklin Institute, the apparatus was now permanently attached to the standard vernier compass. The solar unit, however, was unchanged from the original. Young charged Burt $83 for the compass that would be used by surveyor John Mullett, one of Burt's closest friends.

On September 10, 1836, Mullett became an official witness to a curious but significant event. William Austin assigned to Lucius Lyon half of the rights, title, and interest in his new compass.[97] For $300 Lucius Lyon was entitled to receive "one half of all the benefits and advantages that may hereafter accrue from said invention." Already an enterprising politician and wealthy landowner, Lyon appreciated a good investment when he saw it. Burt, on the other hand, knew his friend could provide much needed influential support.

Chapter VI

TRIAL AND ERROR

Where the willingness is great, the
difficulties cannot be great. [98]

- Nicolo Machiavelli (1469 - 1527)

Iowa Surveys

Following the end of the Black Hawk War in 1832 a large tract of land west of the Mississippi was opened to settlers. By 1836 over 10,000 squatters had occupied the area encompassing the present state of Iowa. In August 1836 the General Land Office Commissioner urged the Surveyor General in Cincinnati to "lose no time . . . in contracting for the survey of all the township lines except the fractionals that would be on the north and west boundaries."[99]

William Austin Burt was instructed to run three lines from which all land in Iowa would be surveyed: the first and second correction lines, and the Fifth Principal Meridian between the two correction lines. The nine-mile section of the meridian south of the first correction line was run by Hervey Parke, while Orson Lyon, brother of Lucius, worked the 14 mile section north of the second correction.

The Surveyor General reminded Burt of his important responsibility:

> You are well aware that the greatest accuracy is required, and much is expected from your long experience and superior skill in the sciences.

The official survey of Iowa began in 1836 with William A. Burt running a "standard line due north to the corner of township 78 and 79" from the northern boundary of the Sacs and Fox Indian reservation. The southern point of the meridian had been established in 1832 by Jenifer T. Sprigg. Burt started in a corn field on the southeast edge of the present city of Durant, in Cedar County, Iowa. The first correction line was run west from the meridian to the Indian cession and east to the Mississippi River. The second correction line was also run. Burt's work on the 72 mile stretch of the Fifth Principal Meridian between the correction lines began November 13, 1836, and took only eight days to complete. He encountered few problems except to report:

> The ground is now frozen so deep that I cannot raise mounds for corners. Shall be under the necessity of going over the lines again as Soon as frost is out of the ground to rais them.[100]

Burt had discovered the surveyors chain varied from the standard measurement in cold weather. To overcome this problem, he warmed the chain by the campfire whenever possible.[101] On Dec. 14, 1836, he was given permission to subdivide townships if severe weather prevented the survey party from running the township lines safely.

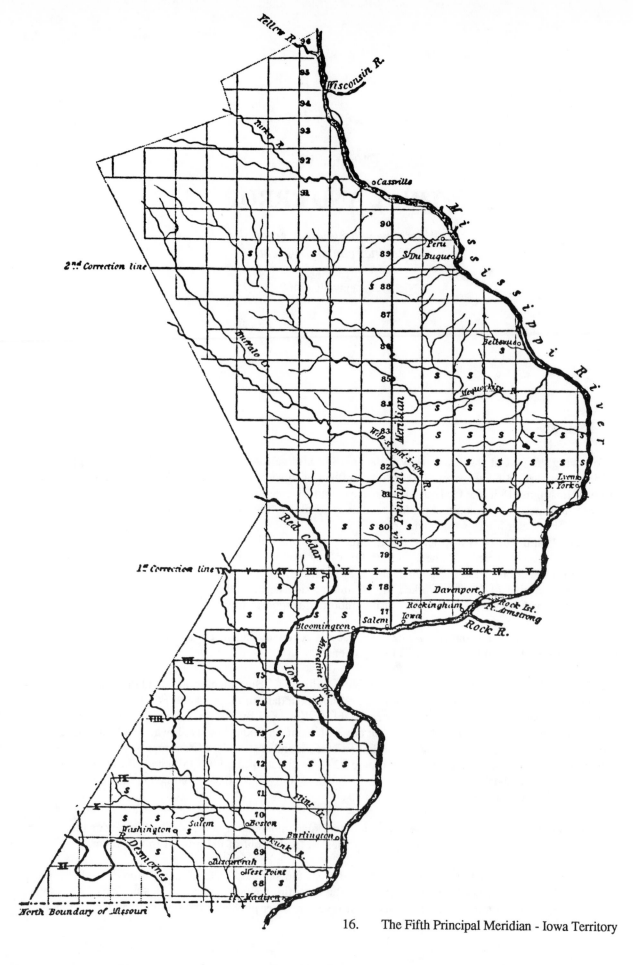

16.　The Fifth Principal Meridian - Iowa Territory

Hervey Parke was less fortunate than Burt. Surveying near an Indian reservation, his only remaining pack horse was stolen, requiring his men to carry all their heavy equipment on their backs. Parke also reported that "one terrible cold night the prarie winds blew out all our fires and our full supply of blankets seemed to afford no protection."[102] In March Parke and his men were nearly blinded by the sun's bright reflection off the snow. When blackening their faces proved unsuccessful, Parke ordered dark (green) glasses that proved helpful.

Nearly a century later one author wrote that Burt was given a unique opportunity which he successfully met. He added:

> It never is, nor will be, correct, to omit from any deed, lease, or easement in Iowa land parcels, reference to the Fifth Principal Meridian. And each repetition of the vital phrase, "west (or east) of the Fifth Principal Meridian, in Iowa" in every land title in our state forever, is an equivalent to saying "W. A. Burt." This approaches nearer to immortality than any other act of any other man of Iowa.[103]

It was also noted that Burt's work on the Fifth Principal Meridian

> is, has been, and will remain well beyond calculation in value for the uses of mankind.[104]

Burt's Variation Compass

It was during this period in 1836 that an accident expanded the role for Burt's variation compass in the U.S. public land surveys. Until then the device had been used primarily to show the variation of the needle, and the magnetic compass was used to run the lines. One day surveyor Harvey Mellen, compassman in Burt's party, accidentally dropped his magnetic compass, bending the sight vanes beyond repair. With no other alternative he resorted to using the variation compass on the lines and found that it was both quicker and more accurate than the magnetic compass. With increased use its value became even more apparent.

Burt proudly informed Lucius Lyon:

> I am now say to you that from mutch experience in the use of the Variation Compass that it exceeds my expectation in point of accuracy, for example I close eleven townships on the last of my work east of the 5th Principal Meridian when the aberations of the needle was grate and the largest difference in closing from what it should be for perfect work was 100 links. inaccuracies in measuring you know is to be expected. . . . Notwithstanding I made a first rate close only 5 links of being parallel to the Township South of it my opinion is that a well made Instrument of this kind in Skillful hand will run a line six miles & not vary from the true corner. Some step must be taken to supply the demand for the Variation Compass & a small book should be published on the use of this instrument.[105]

His reference to a supply problem indicated a real concern. In July 1837 Burt had to inform Morgan Martin, a prominent land developer in Green Bay, Wisconsin, that he had been waiting overtime for three new surveying compasses with a better variation compass attached.[106] Together with land speculators Solomon Juneau and Byron Killbourne, Martin had used the plat based on Burt's 1835 survey to purchase at $1.25 per acre most of the area comprising the business center of modern Milwaukee.[107] Although William Young had agreed to send three units, they did not arrive until August. Young promised three additional compasses would be forthcoming and expected to be able to devote more time to making them in the future.

In May 1837 William Austin and his son Alvin contracted to subdivide 1,341 miles of line west of the Mississippi River. Their use of the new compass during this survey increased William Austin's excitement. In August he notified William Hamilton, Actuary of the Franklin Institute, that the opinion expressed by the subcommittee in December 1835 "has been more than realized by me in practical utility of the instrument." Burt added his opinion that it was "incomparably better than any instrument now in use in the first surveys of the public lands."[108]

Difficulties

The new compass had passed its initial tests, but Burt was not entirely satisfied. Modifications were needed for greater accuracy. As with most new inventions, trial and error was necessary. Correspondence between Burt and Young indicates that occasionally the two creative minds clashed. Once, when the convex lens in the sighting bar was not placed correctly, Young explained, "I was governed as I think I generally should be, by the directions given me."[109] He did, however, accept responsibility for failing to have the declination index low enough to read. Burt was in charge, and Young accepted that fact:

> I hope you will succeed in making such alterations as may make the instrument in every way perfect & to please you, I should be willing to make suggestions, did I exactly know when you wished the alterations, & if any suggestions from me would be agreeable.

Two months later Young wrote that he had a plan to improve the effectiveness of the variation compass, and later he forwarded a detailed sketch to Burt for his comment and approval.[110] Burt acknowledged to Lucius Lyon that Young's plan was impressive, but he chose to ignore it. Instead, he forwarded his own plan to Young the following April. Graciously Young accepted it and told Burt he was "pleased that you have improved your instrument in the manner you speak of."[111] He then invited Burt to Philadelphia to supervise construction of the new improved compass, but the offer was declined. Was Burt upset and possibly considering another manufacturer for his compass? Young, nevertheless, continued to show support for Burt's idea:

> The plan by which you have so far improved and simplified your instrument to produce the results you mention must indeed be ingenious.[112]

In fact, in August 1838 Burt decided to find an additional manufacturer for his variation compass.[113] The most valid reason appears to be Young's inability to keep up with the demand. He could not be faulted, because he would never sacrifice his high standards of quality for mass production. There were only a limited number of competent workers available to do the type of precision work Young demanded. In addition, there were many other orders to fill, as Young's reputation continued to grow.

In 1841, in a letter to all of his deputies, Surveyor General Haines revealed the identity of the new manufacturer of solar compasses:

> Most of these [solar] compasses have been made in this city (Cincinnati) where they may be procured, to order, by addressing the manufacturer, Mr. Henry Ware. His price for the patent solar compass is one hundred dollars, independent of the charge to the patentee, which is from fifteen to twenty dollars.[114]

Haines did not even mention William J. Young who had undoubtedly made more solar compasses than Henry Ware, and charged about the same price for each instrument. In the period between November 1835 and April 1841 Young made the original model for the patent office, two "variation apparatus'," six "variation compasses," and five "solar compasses" for Burt.[115]

It is unknown how many solar compasses Ware actually made, but he built mathematical instruments in Cincinnati from 1839 until at least 1858.[116] Haines may have been mistaken about the royalty fee charged by Ware, since Burt had arranged with Young to receive only a $10 fee.[117]

In April 1841 Young told Burt that he planned to make a solar compass with a telescope attached, but it is unknown whether it was completed.[118] Young apparently did not ship Burt any solar compasses between May 1841 and April 1844. During this period Rasselas Prince Whitcomb, of Cincinnati (New Richmond), was making and selling Burt Solar Compasses, including one sent to Mississippi and another to Louisiana.[119] In 1848 Whitcomb sent Burt a telescope with instructions for mounting it on a solar compass.[120]

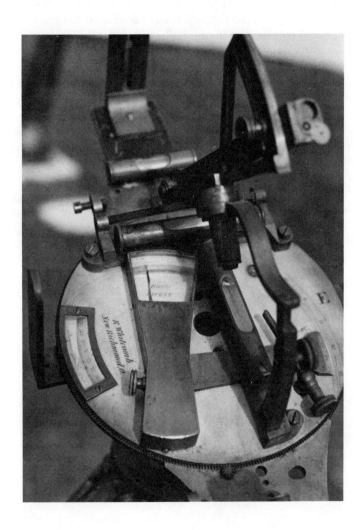

17.   Burt's patent solar compass made by R. Whitcomb (c. 1840's)
courtesy Roy Minnick

William A. Burt played a role in Michigan's ill-fated system of public works during the late 1830's. On April 4, 1838, he became one of the first members of the Board of Commissioners on Internal Improvements, as a representative from Macomb County.

One of the first acts approved by Michigan's State Legislature was to provide for construction of certain works of "internal improvement." On March 21, 1837, the Board of Internal Improvements was authorized to "locate, survey, and construct" three railroad lines and three canals.

The Northern Railroad was to extend from Port Huron to Lake Michigan. Shortly after he was appointed to the board, William Austin made the survey of the southern section of this line from Port Huron to Saginaw City. At the time, however, he was skeptical and considered the massive government project a utopian scheme better managed by private enterprise.[121]

It was an accurate assessment, for the entire program of public works collapsed, and very little was actually accomplished. Part of the blame was attributed to the failure of bonds associated with a $5 million loan used to finance the project. It caused a severe economic hardship for Michigan that proved politically useful to the Whig Party. Their candidate for Governor, William Woodbridge, successfully defeated the incumbent, Stevens T. Mason, who had negotiated the bonds for the "Five Million Dollar" Loan.

While the Central and Southern Railroads were completed with private investment, the main railroad development in Michigan came after the Civil War. Of the canals originally proposed only the canal around the St. Mary's falls ever materialized.

Fifty thousand dollars had been approved during 1837 and 1838 by the state legislature for the "Soo" Canal. In 1839 the governor reported that a contract had been awarded and work on the canal would begin soon. The U.S. government, however, opposed the plan, and military troops were sent to ensure it was not carried out. It was several years later, following William Austin Burt's important discovery in the Upper Peninsula, that plans to build the canal were renewed. The Burt family would be instrumental in bringing about the eventual completion of this vitally important canal.

PART TWO

Chapter VII

ACQUIRING THE UPPER PENINSULA

"Si quaeris peninsulam amoenam":
If you seek a pleasant peninsula look about you.

- State motto of Michigan

Michigan Statehood

William Austin Burt's work in Iowa from 1836 through 1839 occupied him during the crucial period of Michigan's fight for statehood. The final outcome of the struggle, however, profoundly affected his future, as the need for new surveys increased significantly.

Lucius Lyon, the newly elected delegate to Congress, presented Michigan's first formal petition for admission as a state on December 11, 1833. A dispute, however, arose between Michigan and Ohio, centering on the boundary line between the two states. When Ohio was admitted to the Union in 1803 its constitution included a section pertaining to boundaries that included the mouth of the Maumee River. This was an alteration from previous maps, so Congress refused to accept this change in boundaries.

When the Michigan Territory was established, two years later, its southern boundary disregarded the Ohio Constitution. Instead, Michigan acknowledged the boundary established by the Northwest Ordinance of 1787, and authorized the line to be surveyed in 1817. Surveyor General Tiffin, however, a former Ohio governor, instructed that the line be run according to the Ohio Constitution. When Michigan's Governor Cass protested, another line was run, as requested by Congress. The area between the two lines became known as the Toledo Strip, eight miles wide on the east, and five miles wide on the west. When Lucius Lyon presented Michigan's petition for admission to statehood, the dispute was renewed.

In 1835, following bitter debates between the two states, Michigan's Governor Mason led a militia to the Toledo Strip to arrest anyone representing Ohio interests, including nine surveyors re-marking the line. Fortunately, it was a bloodless confrontation.

On June 15, 1836, by a vote of 143 to 50, Congress presented a compromise. Ohio was granted the boundary it claimed, but Michigan received from Wisconsin Territory the western two-thirds of the rugged wilderness now called the Upper Peninsula. Lucius Lyon played an important role in this action, as one of the first politicians to promote Michigan's acquisition of the northern strip of land. He was convinced Michigan would lose the Toledo strip, and reasoned that the Upper Peninsula might someday be valuable. It would also give Michigan a more imposing appearance on the map.

Except for Lyon, however, the reaction in Michigan was overwhelmingly negative. It was six months before the compromise was finally accepted, and on January 26, 1837, Michigan became the Union's 26th state. As a symbolic reminder of its struggle for statehood, the official seal of the state of Michigan, as adopted by the constitutional convention of 1835, retains the date of 1835 rather than 1837, the year of actual admission.

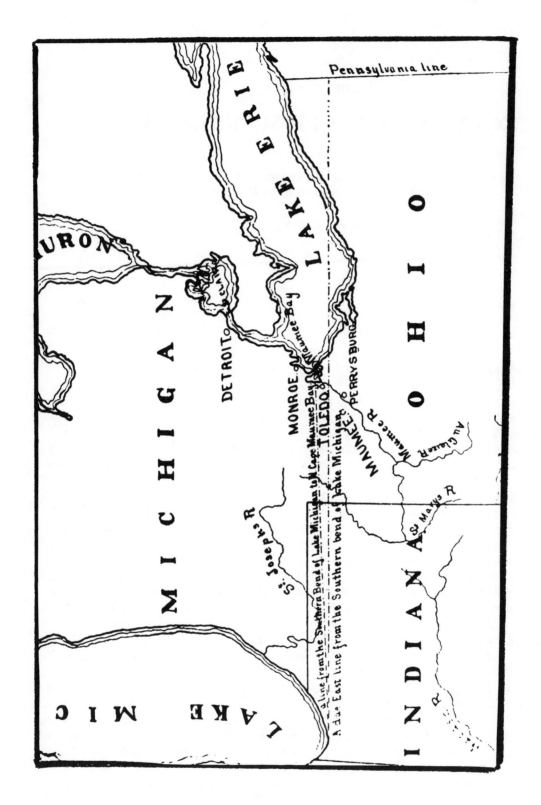

18. The Toledo Strip (*MICHIGAN HISTORY Magazine, 1934, XVIII, 208*)

More Survey's Are Needed

With the Treaty of Washington in 1836, Chippewa and Ottawa Indian lands were ceded to the United States. Now, with Michigan formally admitted to the Union, Surveyor General Robert Lytle was notified by the General Land Office Commissioner to intensify the public surveys. He reported, however, that only one surveyor could be found to run the township lines east of the Michigan Meridian and north of Saginaw Bay. It was not surprising. The Saginaw Bay area had been described by another surveyor as "a gulf of terror":

> [Proceeding northward] the scenery became so wild and forbidding, the country so poor, that a century must certainly elapse before the crowded people of the east, in desperation, would seek homes in this remote section.

According to another report, the heavy forests shut out the sun, and the water courses were choked with logs and brush.[6]

Not everyone found the Saginaw area hostile. The noted French philosopher Alexis de Tocqueville and his companion, Gustave de Beaumont, visited there in 1831. They were advised in Detroit by the territorial land agent that the area was covered by an almost impenetrable forest, "full of nothing but wild beasts and Indians."[7] Undeterred, Tocqueville wrote:

> We could not contain ourselves for joy, for having at last discovered a place to which the torrent of European civilization had not yet come.

Following their tour of several cities, Tocqueville added:

> Everything is extraordinary in America, the social conditions of the inhabitants as well as the laws; but the soil upon which these institutions are founded is more extraordinary than all the rest.[8]

Part of the problem in locating surveyors was that most of the experienced deputies, like William Austin Burt, were engaged in other work. In his annual report to the GLO Commissioner, Surveyor General Lytle requested the contract wage be increased from $3.00 to $4.00 per mile. As justification, he reported that surveyors had never before encountered such an "interminable wilderness, covered with an unbroken forest . . . bounding with burnt and fallen timber, thickets, swamps, marshes, and ponds."[9]

Surveyor Lewis Clason, working an area northwest of Saginaw Bay, sarcastically addressed a letter to Lytle from "Musquito Headquarters." He wrote:

> The musquitoes and gnats . . . are more numerous and more voracious than I have ever met with elsewhere. I do not doubt that their number is sufficient, upon every quarter of an acre, to consume the last drop of blood in my entire company. They swarm about us [like] a cloud, not to guide us through the wilderness, but to devour.[10]

In 1838 Iowa and Wisconsin territories were formed into a new surveying district. Ezekiel S. Haines, new Surveyor General for Ohio, Indiana, and Michigan informed GLO Commissioner Whitcomb that little progress had been made during the year in extending the surveys. The forests were so dense that pack horses could not penetrate them, and surveyors had to pack supplies on their backs.

Nevertheless, both Austin and Alvin Burt accepted contracts to run 660 miles of line for $2.75 per mile. Alvin also contracted for an additional 262 miles at $4.00 per mile in the area north of Saginaw Bay.[11] Appropriations had run out, however, so the work was postponed until spring of 1839.

The harsh difficulties experienced by surveyors who worked in the area northwest of Saginaw Bay during the period 1837 through 1839 cannot be overemphasized. They undoubtedly contributed to the enormous amount of defective and fraudulent work that would be revealed during the following decade, a revelation that would change the lives of many men.

While other surveyors were encountering difficulties in Michigan, William Austin Burt was still busy surveying in Iowa Territory. On October 24, 1839, he was instructed by Surveyor General Ellis, in Dubuque, to run nearly 600 miles of line in the southern portion of the Winnebago Cession. The General Land Office was acutely interested in determining the presence of minerals, and Burt was instructed to report any discoveries of lead, copper, and iron.[12] It was his first effort at geological exploration, but it would not be his last.

Chapter VIII

THE UPPER PENINSULA SURVEYS BEGIN

Progress, therefore, is not an accident,
but a necessity... it is part of nature. [13]

- Herbert Spencer (1820 - 1903)

Burt's Contract

By 1840 the linear surveys had reached the third correction line located in the upper portion of the Lower Peninsula. On January 14th William Austin Burt was given an important contract with instructions to extend a range line from the third correction line, across the Straits of Mackinaw by trigonometrical process into the Upper Peninsula.[14] Additionally, all the township lines in the U.P. east of and including the line between Ranges eight and nine west of the Michigan Meridian were to be run. The work was to be completed by November 1, 1840.

As Burt prepared to begin the linear surveys of the Upper Peninsula in 1840, Michigan's State Geologist Douglass Houghton headed for the Keweenaw Peninsula to extend the geological survey of the U.P. Houghton and other members of the Henry R. Schoolcraft expedition had found traces of copper there in 1831. The geological survey of the Upper Peninsula began in 1839, with Houghton making geological and topographical observations. He reported the area from Sault Ste. Marie, where Burt was headed, to the Menominee River to be a "perfect wilderness."[15]

In June 1840 William Austin informed Surveyor General Haines that he had passed the Straits of Mackinaw, between ranges three and four west, and intersected the Upper Peninsula. It proved to be the most difficult assignment William Austin had ever received. He nearly wore out his "coat and pantaloons," and wrote to Phebe requesting that she make him an outfit of the strongest kind of bedticking.[16]

By now all five of William Austin's sons had become proficient surveyors and were assisting him on this survey. John, at age 26, served as compassman. He would later recall his first glimpse of the Upper Peninsula, but he could hardly have envisioned the significant role he would play in its later development.

At Sault Ste. Marie Burt drove a cedar stake to establish the northern point of the Michigan Meridian. A concrete monument was erected on this site 124 years later to honor this historic event.[17] The bronze plaque on the monument reads:

> On Aug. 25, 1840, U.S. Deputy Surveyor William A. Burt established at this point the north end of the principal meridian from which all land in Michigan is surveyed. This monument erected in 1964 by U.P. Chapter, Michigan Society Registered Land Surveyors.

The Upper Peninsula surveys had begun, and Burt's Solar Compass would play an important role in accurately establishing the township lines.

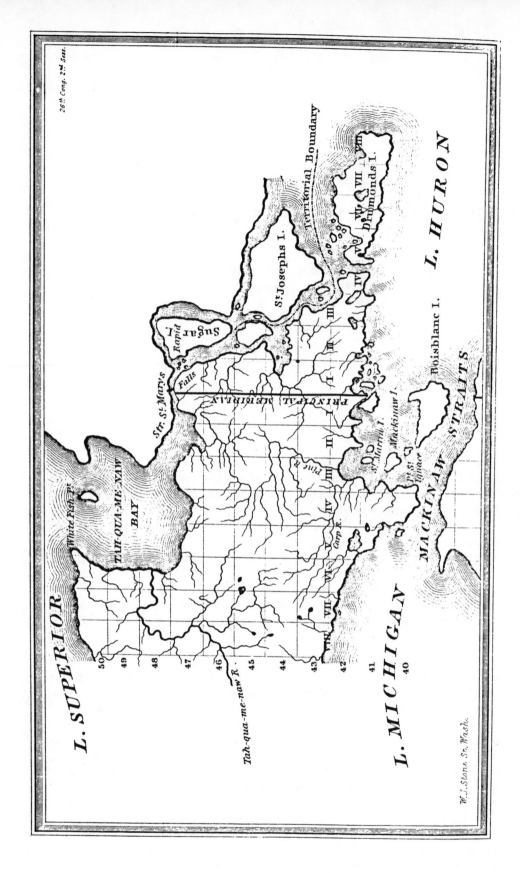

19.   The northern point of the Michigan Meridian as established in 1840
      (SS#377, S. Doc 61, 26-2,SG rept. - Oct 24, 1840)

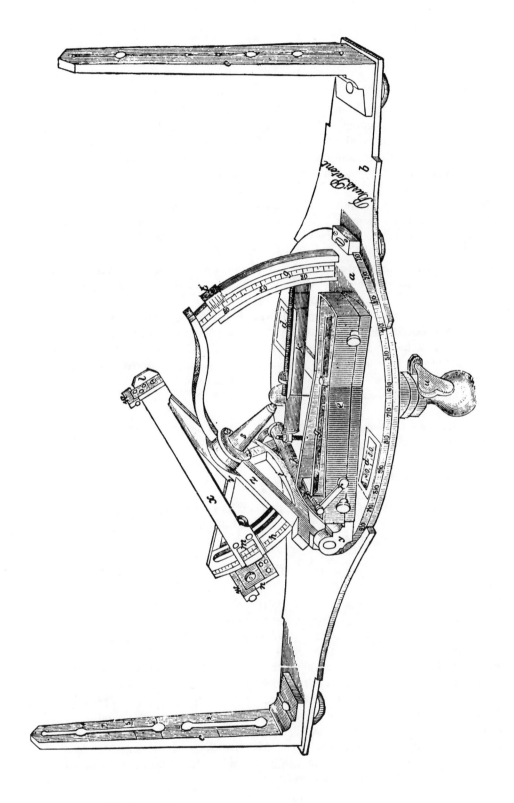

20.   Burt's improved solar compass made by William Young *(Charles R. Tuttle,*
GENERAL HISTORY OF THE STATE OF MICHIGAN (1873), p. 517

Burt's Solar Compass

By December 1840 promotion of Burt's compass for general use in the public land surveys was in full swing. It had been five years since The Franklin Institute's Committee on Science and the Arts had reviewed the original model, so an improved compass, built by William Young, was submitted for evaluation. In its report dated December 14th the committee acknowledged:

> the improvements introduced by its inventor tend to render the instrumentmore simple in its use, and more permanentin its adjustments. . . . It seems to be a very important improvement over the ordinary surveyor's compass, and deserving of great commendation.[18]

For the first time, Burt's invention was referred to as a "solar compass," and it contained most of the features found in later models. In a letter to his son Alvin, William Austin explained the improvements:

> When I left Cincinnati I went on to Washington and from thence to Philadelphia I got Mr. Young to make me an improoved Solar Compass. It differs considerably from the one I had made in Cincinnati. As I now have it, all the adjustments are made by reversals and done in a few minutes thus prooving the Instrument to be in good order. . . . my improooved compass can be used at night on Venus or aney of the planets or aney fixed star within 25° N or S of the Equator.[19]

Burt also informed Alvin that he had contracted with Young "to fill all orders for the Solar Compass." Six of them would be sent to Burt by early spring.

In December 1840, several testimonials supporting the solar compass were obtained by Burt, including one from Surveyor General Haines who wrote:

> I have seen and examined Burt's "Solar Compass". . . and consider it a valuable and important improvement in the surveying compass. . . . I take great pleasure in recommending it to all who feel an interest in the advancement of science.[20]

True to his word, the following year, on April 4, 1841, Haines outlined the advantages of Burt's Solar Compass in a letter to his deputies:[21]

> The advantages of the instrument over every other in use seems to be this: when the sun shines upon it, its adjustment to the true meridian of the place is the work of a moment only, and any course desired may be taken instantly by turning the sights to the same. No matter what the variation of the needle may be, or the strength or variableness of local attraction, the compass gives unerringly the true meridian every time it is set in the light of the sun, and can in no wise be in the least affected by any change of variation of the needle, or any degree of local attraction. This constitutes its superiority over the common compass, particularly for running exterior township lines. Mr. Burt has used it with success for three or four years; and Mr. John Mullett, and his son James, Mr. Hodgman, Mr. Brevort, Mr. Brink, Mr. Mc-Clarnon, and others of my ablest deputies in Michigan, have laid aside the common compass and adopted the solar instrument with high commendation.

On July 3, 1841, Haines' letter was sent by GLO Commissioner Elisha Huntington to all Surveyors General with his own recommendation that Burt's Solar Compass be put to use in all the surveying districts.[22] It was not, however, largely because few knew how to properly use it during the early 1840's.

21.   Burt's Sun Dial made by William J. Young - 1841 *(author's collection)*

Burt's Sun Dial

After Burt completed his 1840 survey, he was invited to Washington, D.C. to confer with the GLO Commissioner about the public land surveys. Highly intelligent, honest, and industrious deputies would be required to meet the increasing need for new surveys. William Austin presented his suggestions for attracting the most qualified surveyors, and many of his proposals, including higher pay, were acted upon.[23]

Before returning home, Burt discussed with William Young his plan for a new type of sun dial that incorporated the principles used in his solar apparatus. He described it as:

> an Instrument to regulate the time for cities & towns, they are fixed permanently on a solid foundation in some convenient place & remain in the same place from year to year. I have ordered one for the city of Detroit.[24]

By mid-April 1841, Young sent Burt two completed "solar dials," charging him $35 for each. Young explained:

> The circles I tried to place on the dials, but after several fruitless attempts to put them as I wished, I gave it up, but as the next best have drawn parallel lines at 5 minutes apart. I did not attempt to try the dials, as it is my intention to make one for my own use, & place it in my yard when it may be subject to examination of the curious.[25]

The sun dials were apparently more of a conversation piece than a popular invention. By 1840 the Swiss watch had achieved world dominance as an accurate timepiece. Nevertheless, at least one sun dial was placed in operation in front of a Detroit jewelry store where it was used by the owner to regulate his clock.[26]

51

Chapter IX

EXTENDING THE UPPER PENINSULA SURVEYS

I conceive it to be of much importance to
Michigan to have the indian title extinguished
to all the land within our state at as early a day as
possible, especially the copper disctrict. [27]

- Wm. A. Burt to Senator Woodbridge, Jan. 3, 1842

New Assignment

Burt's first Upper Peninsula survey in 1840, while historically significant, was not financially rewarding. From the $4.25 per mile he received, William Austin had to pay his crew and provide the necessary equipment and supplies. On one occasion he required additional provisions that cost $2.40. It took two men nearly six days to deliver the items to Burt's camp in the dense wilderness. The added delivery charge was $16.00.[28] To compound the problem, the deputy surveyors were not paid until several months after their field work was completed.

Although he did not complain about it, the financial concern may have accounted for his reluctance to return to the U.P. In a letter sent to his son Alvin in Iowa, he confided:

> I talk of going to Lake Superior surveying again next summer if Congress makes an appro-
> priation for that work. I should much prefer a district in Iowa or Wisconsin if I could get it
> & would commence the work at any time desired, will you talk with the Surveyor General
> about it & let me know the result.[29]

The plan to extend the Upper Peninsula surveys westward to the Chocolate River was presented by Surveyor General Haines to the GLO Commissioner on February 19, 1841. The fourth correction line would be extended both east and west to guide the surveys on the peninsula. Since Haines had decided that all the lines, including section lines, would be run with the solar compass, William Austin Burt was given a contract, on April 1, 1841, to survey about 1,000 miles of township boundary lines in the Upper Peninsula.[30]

Haines decision to exclusively use the solar compass on the U.P. lines created at least one important problem. On April 16th, Burt informed Haines that few surveyors could be found qualified to effectively use the solar compass, so his son John would assist him.[31]

To help meet the increased demand, Young shipped Burt five solars, charging him $100 each. Three months later, possibly in response to the upsurge in usage of the solar compass, Lucius Lyon had his assignment of an undivided one-half interest in the invention officially recorded in the U.S. Patent Office.[32] Lyon, having declined to run for a second term in the U.S. Senate, was active in several business ventures at the time, including an unprofitable investment of $40,000 in the manufacture of salt.[33]

22.   Douglass Houghton (1809 - 1845)
(from portrait by Alvah Bradish- Michigan Historical Collections, v. 39, 1915)

Douglass Houghton

The energetic Dr. Douglass Houghton first came to Michigan in 1830 at the age of 20 to lecture in natural science.[34] He had taught chemistry at Rensselaer Polytechnic Institute in Troy, New York, and was admitted to the practice of medicine in Chautauqua County, New York.

After serving as physician and botanist on the Schoolcraft expedition to Lake Superior in 1831, he established a medical practice in Detroit. Houghton strongly advocated the need for a state geological survey, and soon after Michigan achieved statehood in 1837 he was appointed the first State Geologist.

It was not surprising that on February 1, 1841, in his annual geological report, Douglass Houghton acknowledged the large deposits of copper in Upper Michigan's Keweenaw Peninsula. The earliest French explorers in the area had first heard about the existence of copper from the local Indians. In 1636, the news was communicated to the world in a book published in Paris.[35] A large mass of pure copper, later called the Ontonagon boulder, was found lying on the Ontonagon River bed, in 1765, by Alexander Henry. He established a copper mine in the area, but abandoned the operation after a severe winter.[36]

Houghton first located the two-ton Ontonagon boulder during the Schoolcraft expedition in 1831. After his exploration along the northern slope of the Upper Peninsula in 1840 Houghton reported the area would prove a continual source of wealth to Michigan.[37] He accurately predicted, however, that such wealth would not be quickly attained, and he cautioned "it may prove the ruin of hundreds of adventurers" who were attempting to develop the resources. Despite the warning, the copper rush was well under way by 1843.

Dr. Houghton could only speculate about the existence of iron ore. In 1840, he examined the area west from the Chocolate River to the Michigan boundary line, and reported that ". . . only a small portion of the rocks in this district belong to those which are truly metalliferous."[38]

In his 1841 Geological Report, Houghton wrote:

> . . . so far as regards the ores of lead, iron, manganese and silver . . . I am led to infer that neither of these, unless it be iron, will be so found.[39]

He added:

> Although hematite ore is abundantly disseminated through all the rocks of the metamorphic group, it does not appear in sufficient quantity at any one point that has been examined to be of practical importance.[40]

Appropriations for the geological survey of Michigan that began in 1837 were increased each year until 1841, when they were not renewed. Without adequate financing, the Geological Survey was confined to the western portion of the Upper Peninsula.[41] It was logical that some of the geological work would be assigned to William Austin Burt, who was carrying out the linear surveys in the eastern part of the U.P.

On April 2, 1841, Burt was instructed by Surveyor General Haines to "report . . . any observations which may occur to you about the character of the country."[42] On October 28th, Burt informed Haines that he had made topographical, geological, and weather observations in the Upper Peninsula.[43] Dr. Houghton acknowledged William Austin's work in his Annual Report dated January 25, 1842:

> In the surveys of the upper peninsula east from Chocolate river, I have derived very great assistance from Hon. Wm. A. Burt, who, during the last two years has been engaged in surveying the United States township lines, for through his kindness I have been enabled to locate and determine, much more minutely than could otherwise have been done, the range of several rock formations over a very large district of country.[44]

In effect, Burt's effort was the precursor of the U.S. Geological Survey.[45]

In his 1841 survey notes, William Austin listed the most suitable locations for harbors and the best routes for roads throughout the U.P. He noted that, while settlement in the Upper Peninsula would not be as rapid as in the open prairie to the West, there were many attractions for immigrants, such as valuable fisheries, lumber, sugar groves, and rich mineral wealth. He also predicted:

> . . . it may be reasonably expected that in a few years the northern peninsula of Michigan will become the site of mutch enterprise and speculation and a field for the expenditure of mutch capital.[46]

To support his prediction, early the following year William Austin offered to loan the U.S. Government $4,000 to construct roads in the Upper Peninsula. The U.P. surveys had stimulated considerable interest in the proposals to develop the north with federal funds. Burt suggested to Senator Woodbridge the two best routes, including a military road from Little Bay [De Noc], in Lake Michigan, to Grand Island, in Lake Superior. It would, he believed, "have a greate bearing upon the spedy opening and working the rich copper mines in that region." He added:

Of the importance of the Ste. St. Maries Canal to our state and to the general government I need not speak. perhaps something may be done towards the accomplishment of all of these objects by Congress.[47]

Fraudulent Surveys

The great difficulties encountered by some of the surveyors working the area north of Saginaw Bay in 1837 became even more apparent in 1842. It was alleged that "great imperfections" existed in several of the early surveys. GLO Commissioner Huntington instructed Surveyor General Haines to investigate the earlier work and to fully enforce the instructions of July 1831 which required that "no deputy who failed in his obligation . . . shall be rehired."

On April 11, 1842, William Austin Burt was given a contract to examine the surveys.[48] For his services he would receive a "reasonable compensation", plus expenses. It was discovered that lines in nearly 80 townships were either not surveyed or were so poorly marked that much of the work had to be redone. Burt reported:

> By reference to the foregoing notes it will be seen that the small quantity of subdivisions that have been made are so erroneous that but little dependence can be placed on the surveyed lines or field notes returned to your office.[49]

The actual resurveys of the area were done later by other surveyors, and the solar compass was the required instrument.[50]

The year 1842 marked the beginning of several examinations by the Burts. Austin assisted his father in the Saginaw Bay region, while Alvin was sent by the Surveyor General in Dubuque, Iowa to examine an early survey in Wisconsin Territory. He, too, found inaccurate work and suggested a resurvey.

What is surprising is not that surveyors in those days made errors but that they made so few. The limitations of the magnetic needle compass were nearly overwhelming. While the value of the solar compass was becoming more apparent each year, surveyors, like other professionals, were creatures of habit. Change did not occur overnight.

In 1843 the financial disaster stemming from the Five Million Dollar Loan continued to caused the Michigan Legislature to withhold appropriations for the Geological Survey.[51] The U.S. land surveys in the state came to a halt. Land sales slowed considerably, leaving large quantities of surveyed land unsold.[52] Nevertheless, John Burt, John Mullett, and James H. Mullett all received contracts to resurvey an area north and west of Saginaw Bay. William Austin and Alvin received contracts to survey the Iowa lands ceded by the Sac & Fox Indians.

The Chippewa Indians ceded their remaining land in Michigan west of the Chocolate River (now Marquette County), and the Treaty of La Pointe was proclaimed on March 23, 1843. This opened the way to extending the surveys westward and determining the extent of the mineral resources in that unsurveyed portion of the Upper Peninsula.

Chapter X

A MOUNTAIN OF IRON

Boys, look around and see what you can find!

- Wm. A. Burt, Sept. 19, 1844

Burt's Solar Compass

In early 1844 William Austin completed his manuscript for a booklet entitled "Description of the Solar Compass Together with Directions for its Adjustment and Use."[53] It became the first published work on the subject and was included with each solar compass sold under Burt's authorization. The manual provided surveyors with the best methods of operating and adjusting the solar compass, and it contained several testimonials of support for the instrument, including one written in February 1844 by John Mullett:

> I have used Burt's "Solar Compass" since its first introduction, and can, with confidence recommend it as indispensable to those engaged in running long and standard lines in a new country; such as Indian boundary, state lines, and township lines of the public surveys, &c; and much superior to the common compass, for all surveying purposes. . . .

During the late 1840's some of the approved solar compasses were made by Rasselas Prince Whitcomb in New Richmond, near Cincinnati.[54] He and other instrument makers, including William J. Young, identified each legitimate solar compass with the words "Burt Patent," or "Burt's Patent," clearly marked on the instrument.

There was at least one unauthorized maker of solar compasses. He was Richard Patten, who worked in New York, Washington, and Baltimore. Before 1841, Patten boasted he was "the only Manufacturer of Sextants and Quadrants in New York"; however, many of the instruments that bore his label had been produced by English makers.[55] In 1828 a suit was successfully brought against him for producing and selling a nautical chart that had been copyrighted by another New York instrument maker.[56]

In 1841 Patten planned to make his own solar compasses and explained to the GLO Commissioner that his instruments would be based on principles similar to Burt's, with important improvements.[57] Burt was not overly concerned. In a letter to Lucius Lyon, dated February 29, 1844, he wrote:

> I saw one of Mr. Patten's solar compasses, at Galena [Illinois] last Spring; if all that he has made are like it, he will do no harm, except to impose on those who buy them, for they are of no use; however I think he is a good workman and could be instructed to make them right; probably you had better see him and make some arrangement to have him manufacture solar compasses in a propper manner, and allow you what is right for that privilege.[58]

By proper manner Burt meant that if Patten was going to make solar compasses, he should at least use one of the instruments made by William J. Young as a guide. It is unknown if Lucius Lyon ever discussed this matter with Richard Patten. If he did, it does not appear that Patten ever agreed to pay Burt a royalty.

Alvin Moves to Iowa

On May 9, 1844, William Austin and Phebe received a letter from their son Alvin who informed them that he had moved with his wife and daughter to Iowa Territory. Their new home was in Cascade, a small village southwest of Dubuque, which Alvin and his father had seen and appreciated during their surveys. About the same build as his father, the 28 year old Alvin was readily identified by his distinctive red hair.[59] William Austin was anxious to learn about Alvin's new "prosperity and wellfair," but Phebe had difficulty accepting her son's decision to move so far from home.[60]

It had not been a year of good news for the Burt family. During the previous winter a virus epidemic had swept through Wales Center, New York. Over eighty people perished, including several of the Burts' friends. In Michigan, Austin Burt's son, Horace, barely survived an attack of scarlet fever, but John Burt's daughter, Mary, was not as fortunate. She was seven years old when she died. Wells Burt, suffering from poor health, left Michigan for an extended stay in Rhode Island. For Phebe it was a particularly lonely time as her husband began making plans to resume surveying in the Upper Peninsula.

Discovery

By late 1843 Douglass Houghton had formulated a convincing argument for combining the Michigan geological survey with the U.S. linear surveys. Without such a plan, the geological mapping of the state would be delayed several months, perhaps years, since state funds were not available to finance it. The plan called for Dr. Houghton to coordinate the combined survey that would require the U.S. Government surveyors to make geological observations in addition to their linear responsibilities. They would report on the character of the ledges along the lines, one mile apart, and collect specimens from them.[61] The added cost would be less than one-half cent per acre. Later, trained geologists would follow the surveyors lines and make more detailed observations and reports.

On June 17, 1844, Congress approved Houghton's plan and appropriated $20,900 for the survey "with reference to mines and minerals."[62] On June 25th Houghton signed a contract with the Commissioner of the General Land Office, and William A. Burt was selected as his principal assistant in the field. The survey "with reference to mines and minerals" encompassed 4,000 miles of township and sectional lines in the Upper Peninsula, and it was to be completed within three years.

From Range 23 West, the township lines were to be extended west and north of the fourth correction line, "taking care not to encroach on the probable boundary line of Wisconsin."[63] The northern portion of the boundary between Michigan and Wisconsin would require a new survey before the range lines could properly intersect it. At 52, when most men begin to dream of retirement, William Austin Burt was about to achieve new heights in his career.

Douglass Houghton and William Austin Burt possessed markedly contrasting personalities. Michigan historian John Bartlow Martin wrote:

> Houghton had a sense of history-making which Burt, the tireless surveyor, totally lacked. Houghton was acutely conscious of the romanticizing of earlier explorers, [while] to Burt accuracy was a matter to be assumed.[64]

During July 1844 Burt quickly selected and assembled his crew: William Ives as compassman; Harvey and Richard Mellen, along with James King, to serve as chainmen and axemen; and Jacob Houghton, brother of Douglass, to operate the barometer. Later, two Indians, John Taylor and Bonney, would join the group as packmen. On August 1st the surveyors boarded the steamship *Illinois* in Detroit Harbor. At Mackinaw they rendezvoused with Douglass Houghton, W. Norman McLeod, and the two Indian packmen.

From Mackinaw the men headed west by boat to the mouth of the Escanaba River. Here, Dr. Houghton and McLeod left the survey party and returned to Mackinaw. It was explained that "since the summer of 1844 was almost over . . . [Houghton] could accomplish little except in the line of preparations for the following year's work."[65] Burt did not see Dr. Houghton again until after the season's survey was concluded.[66]

On August 14th William Austin and his crew carried their equipment and supplies several miles along the Escanaba River bank until they reached a sawmill. With lumber he had purchased Burt built a flat-bottom boat, filled it with provisions, and hauled it about 15 miles up river to the south boundary of Township 43 North.[67] Here their survey work began.

Heading north, on the line between Ranges 24 and 25 West, they took accurate measurements, establishing the quarter-section, section, and township corners along the way. On September 15th the line intersected Lake Superior, and Burt spotted the entrance to Presque Isle harbor (near present-day Marquette).[68] The line between Townships 47 and 48 North was then extended westward.

This was rugged wilderness, with thick forests, hills, and swamps. The forest was too dense to permit the use of pack horses, so the men back-packed all the provisions. During the evening of September 18th the surveyors set the post on the township line at the corner between Ranges 26 and 27 West and camped on the edge of Teal Lake.

By the light of the campfire William Austin reviewed his plans for the next day. From the corner post a random line was to be run six miles south to the southeast corner of Township 47 North, Range 27 West (T47N, R27W). The men would then return on the true range line.

Burt had already filled 26 pages of field notes and recorded in his diary several observations made during the survey. He noted that compassman William Ives had been unable to keep up with the group after a thorn became embedded in his leg. Consequently, Burt replaced him with Harvey Mellen. It was the same Harvey Mellen whose accident with the magnetic compass in 1836 revealed the true value of Burt's Solar Compass. Now Mellen was about to take part in the most significant event of his life.

The following morning Burt succinctly wrote:

> East Boundary of Township 47 North, Range 27 West. This line is very extraordinary, on account of the great variations of the needle, and the circumstances attending the survey of it. Commenced in the morning, the 19th of September; weather clear; the variation high and fluctuating. on the first mile, section one.[69]

Jacob Houghton's account was more descriptive:

> On the morning of the 19th of September, 1844, we started to run the line south between ranges 26 and 27. As soon as we reached the hill to the south of the lake the compassman began to notice the fluctuation in the variation of the magnetic needle. We were of course using the solar compass, of which Mr. Burt was the inventor, and I shall never forget the excitement of the old gentleman when viewing the changes of the variationmthe needle not actually traversing alike in any two places. He kept changing his position to take observations, all the time saying, "How would they survey this country without my compass? What could be done here without my compass?" It was the full and complete realization of what he had foreseen when struggling through the first stages of his invention. At length the compassman called for us to come and see a variation which will beat them all. As we looked at the instrument, to our astonishment the north end of the needle was traversing a few degrees to the south of west. Mr. Burt called out, "Boys, look around and see what you can find." We all left the line, some going to the east, some going to the west, and all of us returning with specimens of iron ore mostly gathered from outcrops. This was along the first mile from Teal Lake. We carried out all the specimens we could conveniently.[70]

23.   Burt's survey party camped on the edge of Teal Lake
      (Tuttle - *GENERAL HISTORY OF STATE OF MICHIGAN, 1873*)

Following their discovery of iron ore, Burt and his party quickly resumed surveying. There was much to do before the severe weather forced them to leave the Upper Peninsula. While heading north along the east boundary of T47N, R27W, Burt recorded that the magnetic needle of his compass was deflected by as much as 87 degrees. He added:

> In some places on the North half, the Needle would not take any direction but dip to the bottom of the box. Two good solar compasses were used on this T. Line & the Variation of the Needle determined by both. When the variation was above 45° or 50° the needle appeared to be weak like one nearly destitute of magnetism. Spaltoric and Haemaltic Iron ore abound on this line.[71]

It rained the entire day September 20th, and the following morning the ground was blanketed with six inches of snow.[72] Without sunshine the surveyors were stranded, unable to run the line with the solar compass. The respite, however, gave a limping William Ives the opportunity to catch up with the crew. Although he missed out on one of the greatest mineral discoveries of all time, Ives' work during the 1844 survey was described as "unusually faithful and historically important."[73]

On September 22nd the weather cleared long enough for the surveyors to extend the range line northward along the east side of section 24. The following day, however, thick clouds prevented the sun's rays from reaching the solar compass. Burt was concerned, for their food supply was nearly exhausted. He later wrote:

> I had plenty of [food] about eight miles distant in the direction we were endeavoring to run the line. No one dare risk the needle as a guide or go themselves, it being cloudy weather.[74]

Fortunately the men survived by dining on three porcupines they had captured and roasted, and when the weather cleared they completed their work. It had taken five days to run only six miles of range line.

A total of 11 samples of iron ore were collected and cataloged by Burt during his 1844 survey.[75] Each piece was placed into a small cotton cloth that was marked with a number and the location from which each specimen was taken. They were then referenced in the field notes and eventually turned over to Dr. Houghton.

On a page in William Austin Burt's field notes, Houghton wrote, "This metamorphic region will prove most important for its valuable iron ores . . ." Burt had located the first of the Great Lakes iron ranges, the Marquette range. Nearly a century later the disclosure was called "one of the famous discoveries in America"[76] and "one of the greatest events in the industrial history of the United States."[77] A new economic course for Michigan was set in motion, and soon hundreds of Irish, Cornish, and Swedish workers would leave productive occupations elsewhere to risk the chance for a brighter future in the U.P. mining industry.

Coincidentally, although significant iron deposits were not known to exist in Michigan before Burt's discovery in September 1844, the U.S. Navy launched its first iron-clad warship nearly nine months earlier and named it the U.S.S. Michigan.[78]

Neither Burt nor a single member of his survey crew ever attempted to capitalize on their discovery. Perhaps that is not surprising. Dr. Houghton had warned copper speculators that instant wealth was unlikely, and that profit from iron ore would be even more difficult to obtain. Unlike copper, which existed in a pure state, the heavy iron had to be extracted from the rock. It would also be costly and time-consuming to market the heavy iron ore. In fact, it was 1862 before the Jackson Iron Mine, established on the site of Burt's discovery, returned a dividend to its stockholders.[79]

Burt's enthusiasm for his solar compass was justified, for its value had been fully confirmed. Millions of dollars would be saved as a result of its general use in the U.S. land surveys. For William Austin Burt, this was far greater satisfaction than personal financial gain.

24. Burt's field notes from east boundary of Township 47 North, Range 27 West

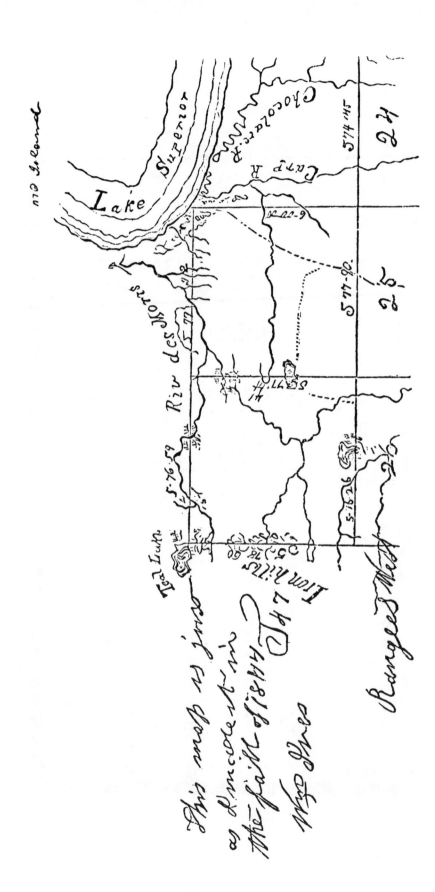

25.   William Ives map of the Iron Hills (*Geological Survey of Michigan - 1873*)

Chapter XI

1845 AND 1846

*Civilization exists by geological consent,
subject to change without notice.* [80]

- Will Durant (1885 - 1981)

Legend of The Pine Stump

The news of Burt's iron discovery was slow to reach the outside world. The copper rush was already underway, and word that gold and silver were also present in the Upper Peninsula attracted the attention of a group of men from Jackson, Michigan.[81]

In June 1845 they formed the Jackson Mining Company, although they knew little about mining or prospecting. The following month, treasurer Philo M. Everett and three companions headed for the Lake Superior region. At Sault Ste. Marie they met a French Canadian Indian named Lewis Nolan who described a mountain of metal near Teal Lake, quite unlike the samples of ore that Everett showed him. It is possible Nolan learned about the ore from Burt's Indian packmen, Taylor and Bonney. Nolan agreed to guide the men to Teal Lake, but he was unable to locate the ore which Burt's party had found the previous year.

In L'Anse, however, they met the Chippewa chief Marji-Gesick who agreed to guide them to the ore south of Teal Lake. Like Nolan, he may have also received his information from a member of Burt's party.[82] Superstition prevented Marji-Gesick from entering the "iron hills," but, on July 23, 1845, he pointed the way for Everett's party.[83] According to local legend they found large chunks of iron, some clinging to the roots of a fallen pine tree. It was located about 3,200 feet west of the range line, in the middle of section 1, T47N, R27W.

The Jackson Mine was soon established, and the city of Negaunee grew up around it. The "Jackson Stump" was formed from the tree containing ore in its roots, and eventually it became a tourist attraction, symbolic of the first iron ore discovery. In 1900 it was reduced to charcoal by vandals, but four years later the Jackson Iron Company replaced it with a large pyramid of native stone that remains as a landmark in the city of Negaunee. A plaque, however, fails to acknowledge Burt's earlier discovery:

THIS MONUMENT WAS ERECTED BY THE JACKSON IRON COMPANY IN OCTOBER 1904, TO MARK THE FIRST DISCOVERY OF IRON ORE IN THE LAKE SUPERIOR REGION. THE EXACT SPOT IS 300 FEET NORTHEASTERLY FROM THIS MONUMENT TO AN IRON POST. THE ORE WAS FOUND UNDER THE ROOTS OF A FALLEN PINE TREE, IN JUNE 1845, BY MARJI GESICK, A CHIEF OF THE CHIPPEWA TRIBE OF INDIANS. THE LAND WAS SECURED BY A MINING "PERMIT" AND THE PROPERTY SUBSEQUENTLY DEVELOPED BY THE JACKSON MINING COMPANY ORGANIZED JULY 23, 1845.

Michigan historian Willis Dunbar, commenting on the plaque, wrote, "The text is inadequate."[84]

A historical marker placed near the Jackson pyramid several years later acknowledges Burt's 1844 discovery, but it credits Everett with locating iron ore in roots of the fallen pine tree in 1845.

26.   Official Seal of the City of Negaunee (*from plate in the author's collection)*

For the Negaunee Centennial (1844-1944), honoring Burt's discovery of iron ore, resident-author R. A. Brotherton presented a different story. Commenting on the fluxuating magnetic needle in Burt's solar compass, Brotherton wrote:

> William Burt and his surveying crew began looking around for the cause of this strange happening. And there, on a little hill to the westward, they find pieces of iron ore clinging to the roots of an over-turned tree - a wind fallen monarch of the forest of which only the stump remains.[85]

The official seal of the City of Negaunee includes a likeness of the pine stump in its center, with the words:

IRON ORE FIRST DISCOVERED IN UPPER PENINSULA OF MICHIGAN AT NEGAUNEE IN ROOTS OF THIS STUMP IN 1844.

The statement refers to Burt's discovery, because Marji-Gesick did not lead Everett's party to the iron ore until 1845.

It is significant that the legend of the pine stump, symbolized on Negaunee's official seal, today acknowledges the discovery of iron by Burt that was overlooked when the Jackson Iron Company erected the Negaunee pyramid in 1904.[86]

Continuing the Upper Peninsula Surveys

In November 1844, following his exhausting survey, William Austin returned home to Macomb County. Three days later, in a letter to his son Alvin, he described the problem he had encountered with the magnetic compass:

> The needle was of no more use than a rye straw would be to run a line with. It would lose any man in a cloudy day that would follow it as a guide.[87]

There was no mention of his valuable iron ore discovery, only satisfaction that his solar compass had accurately guided them through the area. He added, "I was not sure of a correct course only as the instrument was set by the sun." Burt also noted that, because of their late start, only 205 of the 4,000 miles of township and sectional lines had been run during their 1844 survey. They would complete the bulk of the geological-linear survey work during 1845 and 1846.

In February 1845 William Austin and John Mullett were given separate contracts to resume the Upper Peninsula surveys. William Austin's contract with Houghton called for him to run 1,000 miles of township lines.

Burt was concerned about returning to the rugged U.P. wilderness. In a letter to the General Land Office, sent one month earlier, he wrote:

> The thick forests and high trees prevent the sun falling on the solar compass, so I can't prosecute the survey with the same accuracy I had formerly done.[88]

He added that his 1844 survey was his most difficult to date. If Burt's comments were an indirect hint for higher wages, John Mullett was more direct. Although he was offered $5.00 per mile to run the township lines on Drummond Island, Mullett held out for more money. Lucius Lyon, anticipating his appointment as Surveyor General, had already informed Mullett that he intended to hire only deputies who were also good practical geologists, and that meant extra work.[89] Mullett accepted the contract only after Houghton agreed to pay him the difference between what he wanted and what Congress had allocated.[90]

In April 1845 William Austin wrote William J. Young in Philadelphia, requesting a rush shipment of one solar compass for his son John, who was subdividing a district north of Saginaw Bay.[91] A few months earlier Young had taken the initiative to alter his usual method of making a solar compass, and William Austin was not pleased. Young assured him, however, that future solar compasses would confirm strictly to Burt's plan.

Alvin Burt was particularly upset by the changes, but his father reminded him that William J. Young made excellent solar compasses, and "it would take considerable time and expense to find another instrument maker as well instructed in them at present as he is."[92]

On June 23, 1845, Lucius Lyon was appointed Surveyor General of Michigan. One year earlier Lyon had written President Tyler urging that the office of Surveyor General for Michigan, Ohio, and Indiana be moved from Cincinnati to Detroit.[93] Lyon reasoned that most of the unsurveyed lands within the district were in Michigan, and supervision could be improved by operating from Detroit. No action was taken until James Polk, Tyler's successor, offered Lyon the job. He declined unless the office was moved to Detroit. Tyler then received Congressional approval for the move in June 1845, and Lyon accepted the appointment.[94]

In June William Austin returned to the Upper Peninsula to extend the lines of 76 townships north and west along the coast of Lake Superior to Keweenaw Point, at the northern tip of Michigan. Neither Burt nor Douglass Houghton had as yet been paid for their work the previous summer.[95] Houghton began his survey by examining the townships just south of the area in which Burt found iron deposits in 1844. Impressed, Houghton wrote:

> This bed of iron ore will compare favorably, both for extent and quality, with any known in our country.[96]

By early October 1845 Houghton had surveyed nearly 40 townships. He had worked late into the season and hoped to complete just one more township before returning home. On the morning of October 13th there was a brief snowfall, but the clouds soon cleared. Later in the day, however, a severe storm developed as Dr. Houghton and two of his assistants were attempting to row the short distance from Eagle Harbor to Eagle River.

Anxious to complete his work, Houghton ignored the warning from his men to go ashore, and pressed on. It was a fatal mistake, for their small boat soon capsized in a violent gale, sending Houghton to his death.[97] His body was not recovered until the following spring. It was a tragic loss for Michigan and the combined geological-linear survey. Fortunately, most of Houghton's field notes were saved.[98]

At the request of the administrators of Dr. Houghton's estate, William A. Burt and geologist Bela Hubbard completed Houghton's geological report for 1845. In addition to his geological work, Hubbard held several land investments and operated a real estate office in Detroit. It has been reported that Hubbard had arranged with several investors to advise them of any choice mining locations he sighted while working with Houghton in 1845. In return he would receive shares in the new company.[99]

In his 1845 geological report Burt identified the types of rocks found in the copper country and described the topography of the area. He correctly predicted a canal would be built from Portage Lake into Lake Superior, and speculated that the site of the present city of Houghton would "probably become a place of considerable importance."[100]

More Discoveries

Following Dr. Houghton's death, many people were concerned about the future of the combined linear-geological survey. William Austin hoped the plan would continue, and predicted "it will be found to be the cheapest and best yet devised for the public interest."[101]

On August 10, 1846, the geological surveys were extended to Wisconsin and Iowa Territories when Congress provided $30,000 for surveying the "copper region of Michigan, Wisconsin, and Iowa, with reference to mines and minerals." All but $5,000 was apportioned to the Michigan district.[102] The unfinished portion of Dr. Houghton's 1844 contract was assigned to topographical surveyors Sylvester and Hiram Higgins and geologist Bela Hubbard.

Under a contract dated September 7, 1846, William Austin and his sons, John and Austin, ran the exterior lines of 86 townships in the Upper Peninsula, north of the Menominee River. They also extended the fourth correction line west from Range 23 West to the Menominee River. During the season they located an additional 14 beds of iron, including the western deposits of the Menominee iron range.[103]

William Austin estimated that they had seen only about one-seventh of the actual number of iron ore beds, and that northern Michigan's iron region "far exceeds any other portion of the United States in the abundance and good quality of its Iron ores."[104] He also stressed the need to build roads from the iron region to Keweenaw Bay so the mineral wealth could be developed. The granite or greenstone found abundantly in the area could be used for their construction.

After returning to his Macomb County home in August 1846, William Austin organized his geological and topographical notes for Surveyor General Lyon. His report for 1846 and accompanying maps were published later in the U.S. Senate Documents.[105]

According to Burt, the survey had been "interesting from a geological point of view," but it had been unusually difficult and expensive.[106] One packman was required for each member of the survey party to effectively traverse the numerous swamps they encountered. A reporter for the *Lake Superior News* described similar difficulties encountered by another group of Government surveyors running township lines in the Upper Peninsula during 1846:

> In enterprises of this sort, it is only by physical energy, and great powers of endurance that the contractor can realize anything from the prices allowed by the Government from its original surveys. They provision themselves by carrying all on their backs, from depots on the shore. The thickets through which they pursue their work, week after week, and month after month would be declared absolutely impracticable to a person not trained in that school, especially in the vicinity of the lake. No beast of burden could pass without bridges, even in case a pathway should be cut through the matted evergreens that cover the ground. To make a path for horse or mule, would consume more time and labor per mile than the survey itself.[107]

The Burt's Chalets

It was during the period 1844 through 1846 that William Austin and three of his sons (Austin, Wells, and William) moved their families to new homes they had built in Mt. Vernon, located about one and one-half miles north of the original Burt homestead established in 1824. Later known as "Burts chalets", they were described as "easily distinguished . . . by their unusual character and design."[108]

The Wells Burt house, located next to his father's home, is a beautiful octagonal structure that originally included an upper veranda, with space-efficient bedrooms on both the first and second floor. According to the present owners, the name "Burt" is etched on the interior structure of the cellar.

Although the Burts lived in the chalets for only a decade, this was their home during an important period of their lives. All of the houses, except son William's, have survived into the 1980's.[109] Across the street from the William Austin Burt house the Baptist church he helped to establish still maintains an active congregation.

Alvin

William Austin felt fortunate to have spent the 1846 season surveying with his sons, John and Austin. He was concerned, however, about his son Alvin in Iowa Territory, who had been ill for several months. In a letter to his father sent in October 1845 Alvin said that he had been too sick to accept any contracts, although he had not lost his interest in surveying. When William Austin inquired about his desire to become the Surveyor General for Iowa, Alvin replied:

> I have never thought or dreamed of aspiring to the Office of Surveyor General. I think that I could do the duties . . . and would like the appointment. But I am too obscure a person to expect it. Our delegates influence will be in favor of Jones.[110]

It is not surprising that George W. Jones was the leading candidate for Surveyor General of Iowa, since he had held that position from February 1840 to April 1841.

In 1846, shortly after returning home from the copper country, William Austin learned that Alvin's health had continued to deteriorate through the summer. Since early winter, nearly 25 percent of the residents in Cascade, Iowa, had died from respiratory complications. By August, it took all the strength Alvin had just to walk to his mill to grind a small grist. On October 20th, at the age of 30, Alvin suffered "congestive chills" and passed away. It was a sad conclusion to his brave struggle with the illness. His last words to his daughter were, "God bless you, Mary," as she begged him not to leave.[111]

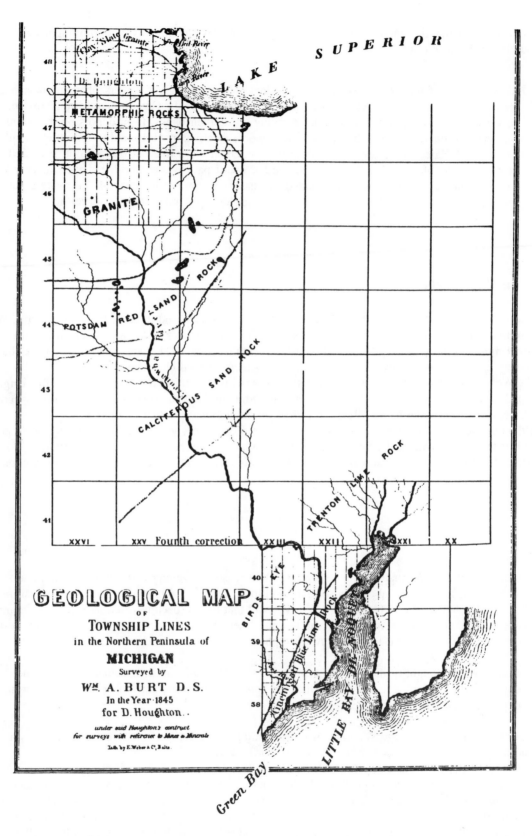

27. William Austin Burt geological map - 1845
(*Surv. Gen. Reports*, U.S. Cong., SS#476, 29-1, S. Doc. 357)

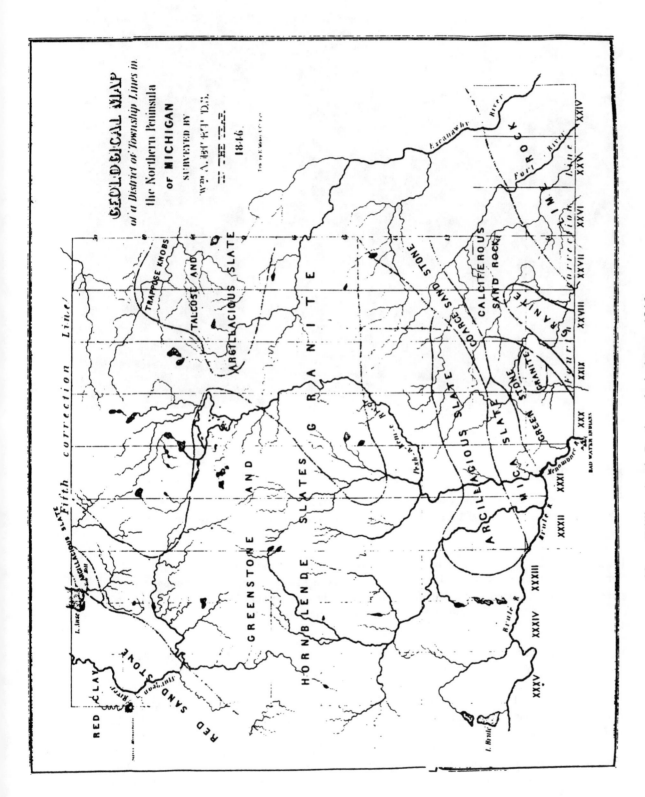

28. William Austin Burt's geological map - 1846
(Geological reports, U.S. Cong., SS#551, 31-1, S. Doc. 1)

69

29.a   William A. Burt's chalet - 1909 *(author file)*

The Solar Compass

If there had been any doubts concerning the value of the solar compass they were dispelled during the Upper Peninsula surveys. Geologist Bela Hubbard, commenting on the results of their work in 1845, wrote:

> This accuracy has been attained by the exclusive use, by all the parties, of "Burt's Solar Compass", an instrument too well known to need more than a bare allusion, but the great value of which has been more than fully confirmed during the surveys of the past season. . . . It seems difficult to imagine how the lands could have been run with the ordinary compass.[112]

Surveyor General Lucius Lyon was equally emphatic in his 1845 Annual Report to the G.L.O. Commissioner. He praised William A. Burt as:

> one of the ablest and most experienced deputies of this office, a good practical geologist, and inventor of the solar compassman invaluable surveying instrument without the aid of which, his surveys in the mining region could not have been made; for the aberrations of the needle are so great on and near the mineral ranges, that its direction will frequently vary twenty, fifty, and sometimes one hundred and sixty degrees in running a quarter of a mile. . . . The solar compass has been used with great satisfaction in all the surveys of the public lands of [the Upper Peninsula of Michigan] for some years past, and its introduction into general use would unquestionably greatly promote the accuracy of the public surveys in all parts of the United States.[113]

29.b   William A. Burt's chalet - 1982 *courtesy Jim Burt*

Lyon's strong support of Burt's Solar Compass, while he was Surveyor General of Michigan, could be interpreted as a conflict of interest. Although he held an equal share with Burt in the patent rights to the instrument, it is possible that this information was known to Lyon's superiors. The modern-day disclosure of investments by government officials was not practiced in the 1840's. Burt's Solar Compass was the required instrument on the Upper Peninsula lines five years before Lyon took office as Surveyor General. Any positive comments Lyon made regarding the solar compass were well substantiated by others.

By mid-1846 the demand for solar compasses exceeded the supply. Burt estimated that he could have sold 30 to 40 of the instruments, but William J. Young could only supply Burt with six units.[114]

Young found himself in a quandary. Burt was continually suggesting improvements to upgrade the solar compass, so it was not practical for Young to stockpile them. Occasionally Burt was dissatisfied with the finished product and Young had to make corrections.[115] It was also a matter of economics. According to Young, it was less profitable to make the solar compass than his surveyor's transit. Young charged $130 for either instrument, but it took much more time and effort to build a solar compass.[116] He confided to Lucius Lyon that he would have to find a more economical way to manufacture solars before he could make any more of them.[117]

71

Apr. 15 - 1909
Wells Burt

30.a   Chalet of Wells Burt - 1902 *(author file)*

In December 1846, Young sent Surveyor General Lucius Lyon two solar compasses for William Austin, which he described as "different from any heretofore sent." He added:

> Adjusting screws will be found to adjust the sight parallel to the meridian and also adjusting screws to adjust the tube so that it will revolve at right angles to its axis—The plan for cutting the vertical and horizontal line of sight, you will find to be novel, but good.[118]

While adjusting screws were included on later versions of the solar compass, Young's plan to position the sights on the side of the compass was not acceptable to Burt.[119]

Throughout the first decade in which William J. Young made Burt Solar Compasses, William Austin maintained complete confidence in Young's abilities as a craftsman. From 1844 through 1846, however, a few of Young's solar compasses were substandard.[120] Young later admitted that during this period he had delegated most of his work to others while devoting his attention to an unsuccessful venture in the iron foundry business.[121] He promised Burt that in the future he would concentrate his efforts to make top quality solar compasses.

30.b   Chalet of William Burt - 1902 *(author file)*

30.c   Chalet of Austin Burt - 1902 *(author file)*

31.   Assumed Michigan-Wisconsin boundary - 1837 (*Michigan Pioneer Historical Collections (1896) XXVII, 387*)

Chapter XII

1847 AND 1848

*...I have surveyed, marked, and
established so much of the boundary
line between Michigan and Wisconsin as
lies between the source of the Brule River
and the source of the Montreal River.*

-- Wm. A. Burt, Nov. 10, 1847

The Michigan-Wisconsin Boundary Survey

It had been nearly seven years since William Austin Burt first crossed the Straits of Mackinaw into Michigan's Upper Peninsula to begin laying out the exterior township lines. In April 1847 the final contracts were issued to complete the work, and on April 27th Burt and his son Austin were assigned a district of 87 townships, bordered on the west and the south by Wisconsin Territory. They received $8.00 per mile, nearly double the rate Burt was paid for his 1840 survey.

On the same date, William Austin received an additional contract to "survey, mark, and designate" the portion of the boundary line between the state of Michigan and Wisconsin Territory that lies between the source of the Brule River and the source of the Montreal River.[122] The work was authorized by the "Act to enable the people of Wisconsin Territory to form a constitution and State Government, and for the admission of such State into the Union."[123]

When Michigan was granted statehood in 1837 its western boundary was believed to follow a continuous water path, from Lake Superior southeast to a point in Lake Michigan.[124] This misconception was corrected in 1840 when the War Department sent Captain Thomas Cram, a U.S. Topographical Engineer, to mark the boundary line. He reported that the two important links in the boundary, the Montreal and Brule Rivers, were not connected. Although he recommended a straight line be run between them, a lack of funds caused the project to be discontinued. Six years later, on August 10, 1846, Congress authorized $1,000 to complete the work.

Lucius Lyon was then directed by the GLO Commissioner to "employ one of your most experienced and competent deputies" for the job.[125] William Austin Burt, then aged 55, was selected for one of his most important assignments.

Burt and his sons, John and Austin, each formed a survey party of about 10 men. On May 14, 1847, the surveyors left Detroit on the steamship *Sam Ward* and traveled to Sault Ste. Marie. Four days later they took the steamer *Independence* to L'Anse, on Keweenaw Bay, arriving on May 23rd.

William Austin was a strong leader who never asked his men to do work he would not do himself. Shortly after they arrived at L'Anse, the surveyors loaded their packs with supplies and climbed a steep hill that ascended from the shore of Lake Superior. The compassman on Austin's survey crew, E. C. Martin, recalled:

> Judge Burt was the first man who shouldered a one hundred pound bag of flour and started
> up the hill; he did not stop until he laid it down at the top. Of course that was an invitation
> for the boys to follow their leader, but some of us found it pretty hard work.[126]

32.   Tamarack tree marked by William A. Burt on June 7, 1847
courtesy Marquette County Historical Society

While John prepared to run the township lines, Austin and his father led their survey crews about 50 miles inland, along an Indian trail, to "Lac Vieux Desert," The Lake of the Desert. After placing a post on the east shore of the lake, a random line was run southeast to Lake Brule where the boundary line was to begin. This trial line was lightly marked to enable the packmen to follow with their supplies. Additional posts were set at each mile and half-mile interval.[127] Rain and heavy snowfall slowed their progress. At several swamps and rivers the men had to build bridges for the pack horses to cross.

Camped on the southern shore of Lake Brule, Burt's party spent a sleepless first night, with the sounds from Indian drums reverberating until dawn the next morning. Burt assured his men that the Indians were not hostile, although he could not have been certain of that. In fact, through the first rays of early sunlight the surveyors were startled to see a large group of Indians surrounding their camp. The chief approached William Austin and presented him with a paper wrapped in birch bark. It was a treaty signed by Captain Cram in 1840. Cram had promised that the next group of U.S. government workers passing through the area would bring gifts to the Indians, but no one had mentioned this commitment to Burt.

Through an interpreter, William Austin reminded the Indian chief that the surveyors were not armed. Since the U.S. Government had purchased their land he had as much right to be there as they had. If the Indians interfered with the surveyors work, it was possible that their "Great Father at Washington" might relocate them to an area west of the Mississippi River. Burt then made his own treaty. If the Indians would allow the surveyors to continue their work, he would share his supplies with them. The treaty was accepted, and there were no more encounters with the Indians.

Burt began his actual boundary survey at a point on the southeast side of a small cove at Lake Brule.[128] He located a tamarack tree on the left bank of the river and marked, "W. A. Burt, June 7, 1847."[129] On July 23, 1964, the location was dedicated as the "State Line Historical Site." The site includes Mile Post Zero, the starting point of the boundary marked by Burt. There is also a Treaty Tree, representing the tamarack tree where Burt and Cram made their treaties with the Indians.[130]

A direct line was then carried to a point in the center of the Lake of the Desert, and from there to the headwaters of the Montreal River where a cedar post was placed. On the northeast side of the post the word "Michigan" was marked, and on the opposite side the word "Wisconsin" was carved. On an adjacent side are the words "State Boundary."

On July 5, 1847, Burt completed his work on the Michigan-Wisconsin boundary line which measured 64 miles, 24.72 chains in length.[131] For his efforts he received $1,000. At $16.00 per mile, it was the highest rate ever received by a public land surveyor in Michigan.

The Michigan-Wisconsin boundary line was the first state boundary completely run with Burt's solar compass.[132] When clouds prevented the use of the solar instrument, the surveyors often spent the the day fishing in a nearby lake.[133] In 1851, to optimize use of the solar compass during fair weather, Surveyor General Noble hired assistant deputies or "random line surveyors," to assist the contracting surveyors. This practice, however, was discontinued when the GLO Commissioner required each contract deputy "to do the work in his own person."[134]

A cadastral engineer who resurveyed Burt's line in 1929, reported it was "one of the best early surveys which he had re-established."[135] He located 75 percent of Burt's original mile and half-mile posts, and noted that their position varied only slightly from his own bearings.

Burt and his party faced many hardships during their 1847 boundary survey, but they recorded no serious complaints. Surveyor Harry A. Wiltse, however, was more candid. He had worked on the Fourth Principal Meridian and the third correction line in Wisconsin during 1847, and he described the "dreadful swamps through which we waded," "the rapid creeks and rivers that we crossed," and "the swarms or rather clouds of mosquitoes, and still more bothersome insects" which constantly plagued them. He also mentioned having to go three days without food, and added, "you can form some idea of our suffering condition." He concluded:

> I . . . would not again, after a lifetime of experience in the field, and a great fondness for camp life, enter upon the same, or a similar survey, at any price whatever.[136]

Closing the Township Lines

With the survey of the Michigan-Wisconsin boundary line completed, William Austin and his sons worked together to extend the township lines northward from the boundary to the fifth correction line. By mid-July 1847 the three survey crews of William A., John, and Austin Burt were nearly out of food and other provisions. Fortunately, packers for John and Austin located a crew of geologists working near the Ontonagon River, with plenty of food to spare.[137] It was a close call. Some of the men in William Austin's crew were on the verge of starvation. When the last of their provisions was gone, William Austin confidently assured his men that they would have food and fresh supplies by nightfall. It was an accurate prediction, for the packers arrived shortly before sunset.[138]

On July 22nd the three survey crews rendezvoused at a geologists' camp near the fifth correction line. Wells Burt was among the surveyors. The geologists were assistants of Charles T. Jackson, who had been sent to the Upper Peninsula to classify the land for mineral and agricultural purposes. That evening geologist W. Gibbs recorded in his field notes:

> . . . we were in the midst of a pretty large assemblage, including Mr. Higgins' packmen. Judge Burt gave us much information with regard to the country he had traversed this season and the last.[139]

77

33.   Completed Michigan-Wisconsin boundary line
(Michigan Pioneer Historical Collections, 1896, XXVII, 389)

The Burts hoped to complete their contract by the end of September, but geologist Gibbs informed William Austin that he planned to continue his work in the copper country until late October.[140] With the fate of Dr. Houghton still fresh in his memory, Burt warned Gibbs that treacherous weather could be expected any time after September. He was right. While trying to return home, on October 11th, Gibbs was stranded at Copper Harbor for three weeks by severe gales and snow storms. It was mid-November before he arrived at his home in Boston.

Lucius Lyon's Concern

Although the task of laying out Michigan's township boundaries was nearly completed by the end of the 1847 season, Surveyor General Lucius Lyon was less than optimistic. He had received several complaints from settlers that many of the original surveys in Lower Michigan were defective. In his 1847 Annual Report to the Commissioner, Lyon wrote:

> The more the old surveys in the northern part of [the Lower P]eninsula are examined, the more certain it appears that a large portion of them have been so loosely and fraudulently made, and that extensive resurveys will be necessary.[141]

In June 1847 the experienced John Mullett had contracted to survey the township boundaries of a large district between Green Bay and the Menominee River. Unfortunately, according to Mullett, "the swamps were too numerous, extensive, and difficult ."[142] He explained that his physical health simply could not take it. Lyon subsequently deputized a young surveyor, recommended by Mullett, to complete the job. The following season, on April 22, 1848, when Mullett contracted to subdivide 15 townships within the area, he agreed to examine the young surveyor's work. To his dismay, Mullett found that many of the surveyor's lines were carelessly run and his field notes contained more fiction than fact. Embarrassed, Mullett sent his son James to resurvey the area.

Lyon's immediate concern, however, was to complete the job of subdividing Upper Michigan's mineral districts. On April 22, 1847, Commissioner Young instructed Lyon to separate the survey "with reference to mines and minerals" from the linear surveys. Unfortunately, no competent linear surveyor was willing to work in that area for the $4.50 per mile maximum price allowed. As a result, Lyon recommended to the land Commissioner that the price be raised to $6.00 per mile.[143]

By May 1848 Congress had not yet approved the increase in appropriations, but Lyon was unwilling to delay the surveys for another year. Since each of William Austin's four remaining sons had agreed to work for the higher price, Lyon sent them into the Upper Peninsula without a formal contract. John and Wells ran section lines in the iron region, and later they received the full $6.00 per mile. Austin and William subdivided the townships they had outlined the previous season. Since their work was closer to the shore of Lake Superior, they received only $5.00 per mile after the contracts were officially drawn up on December 9, 1848.

These were the first of several contracts that William Austin's sons Wells and William would receive for work on the U.S. public land surveys. During 1848 John discovered several more deposits of iron, including the location where the Republic mine was established in 1870. John said "the attractive power of this body of ore affected the needle for a distance of 6 miles."[144] The Republic mine, purchased by the Cleveland-Cliffs Iron Company in 1913, was still operating in the 1980's.

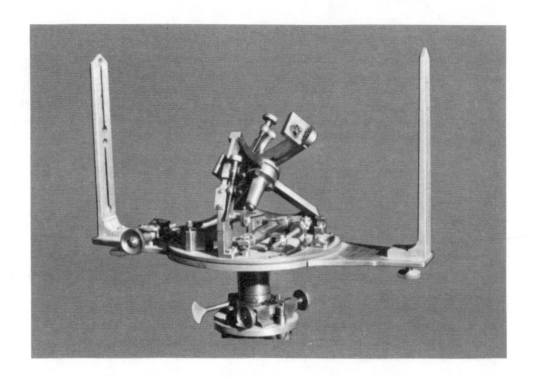

34. Aluminum Burt Solar compass (c. 1880's) (*courtesy Paul Herndon - BLM) Fall 1977, OUR PUBLIC LANDS, p. 4.)* This compass is now attached to the Eastern States Office, BLM, Silver Spring, Maryland

Improving the Solar Compass

During 1848 William Austin kept busy with several projects at his Macomb County home. In February, he sent William J. Young his plan for making more accurate adjustments to his solar compass.[145] It included a special brass "adjuster" to enable the surveyor to quickly and easily adjust the lenses and silver plates on the compass. In April Young sent Burt four solar compasses, with the new adjuster included with each instrument.[146]

William Austin also asked Young about the possibility of reducing the weight of the solar compass. Young suggested that German silver would be much better and lighter than brass, but it would also be more expensive.

The solar compasses continued to be made of brass or bronze until 1880, when some of them were made of aluminum. One such instrument was manufactured by W & L.E. Gurley, in Troy, New York. It was one of the first industrial uses of aluminum in the United States.[147]

At $130 each the high cost of the solar compass continued to be a problem for some surveyors. A common magnetic compass could be purchased for $51.[148] Consequently, because of the greater expense most surveyors continued to use the magnetic compass on the U.S. land surveys in districts where the solar compass was not required.

Clearly the cost problem had to be overcome if Burt's Solar Compass was to be widely accepted by the contract surveyors. The solution, however, would not be easy, because William J. Young was barely making a profit at the price he was charging for them. Nor was Burt getting rich from sales of his invention. Although he had originally contracted with Young to receive a $10 royalty fee, Burt usually waived his fee to encourage more surveyors to purchase the instrument.[149] It is estimated Burt received less than $100 in royalty payments during the 14 year term of his patent.

Chapter XIII

1849

*A faithful and good servant is a real
godsend; but truly 'tis a rare bird
in the land* [150]

- Martin Luther (1483 - 1546)

Political Changes

During 1849, political changes occurred that would indirectly affect the public land surveys. In January, Zachary Taylor, a national hero in the Mexican-American war, took office as the 12th U.S. President. Despite his political inexperience, Taylor was nominated by the Whig party based on his grassroots popularity. He defeated Michigan's Lewis Cass, the Democratic candidate, by a mere 36 electoral votes, aided by the Free-Soil candidate Martin Van Buren who pulled considerable support from Cass.

On March 3, 1849, the Department of the Interior was established by Congress to represent domestic interests of the United States. It comprised several independent government bureaus, such as the U.S. Census Bureau and the U.S. Patent Office. Other bureaus, including the General Land Office, were transferred from other departments.

The GLO transfer from the Treasury Department reflected a significant change in emphasis. The discovery of California gold in January 1848 made expansion to the West a top priority. The need for new surveys quickly accelerated, while revenue from the sale of land assumed a secondary importance.

In July 1849 a new Commissioner of the General Land Office was appointed, ending Abraham Lincoln's strong bid for the position. Lincoln, a staunch supporter of Zachary Taylor for President, considered the job a political plum.[151] He confidently predicted he would become the new Commissioner, but Justin Butterfield, another prominent Illinois lawyer, was chosen instead. Bitterly disappointed, Lincoln called Butterfield's appointment "an egregious political blunder."[152]

Within five months after taking office, the new GLO Commissioner would confer with William Austin Burt in Washington, D.C. to discuss the future use of the solar compass in the expanding public land surveys. A new era was about to begin.

A Matter of Conscience

In Michigan mounting pressure from mineral and lumber interests caused Surveyor General Lucius Lyon to intensify efforts to complete the Upper Peninsula surveys. Nearly all of his deputies, including three of William Austin's sons, accepted contracts for the work. John surveyed in the iron region while William subdivided in Baraga County near L'Anse. Wells' district was just west of Whitefish Bay in the eastern part of the U.P.

In the Lower Peninsula, Lyon faced the problem of correcting several districts in which faulty surveys had been reported. Although Congress had already appropriated funds for resurveys, Lyon first wanted a full estimate of the problem before proceeding with the work.[153] On April 5, 1849, William Austin

Burt and surveyor Orange Risdon were hired to examine the original surveys of 280 townships in the northern part of the Lower Peninsula. William Austin was responsible for investigating 161 of the townships situated between Saginaw Bay and Grand Traverse Bay, while Risdon would work the area south of present-day Traverse City to Muskegon.

The assignment would severely test the character of both men, because many of the surveyors whose work they would examine were close friends or associates. Even family members were involved—lines run by Alvin and Austin Burt in 1838 would also be inspected.

On April 16, 1849, Burt's survey party left Macomb County for Saginaw. The men did not return home until September 23rd.[154] During those five months the extent of irregularities uncovered sent shock waves throughout Michigan and the General Land Office. Burt found that a large portion of the field notes from surveys in 91 townships were fictitious. He was particularly distressed to discover that his friend Henry Nicholson had committed classic fraud on an 1837 contract that had been guaranteed by Burt and John Mullett.[155] Nearly half of the lines Nicholson described in his field notes had never been run. Further, his work performed in 1838 was found to be equally fraudulent.

Gross inaccuracies were uncovered in other districts. It was unpleasant for William Austin to report these revelations, involving his friends as they did, but it was a matter of conscience. Although surveyor John Mullett was innocent of any impropriety, he suffered as much as anyone by the reports of fraudulent work, for he had been surety on four of the eight contracts in districts found to be defective. In an area assigned to Mullett's son James, Burt found many of the lines had never been run.[156] According to Burt's chainmen the few that had been were "so grossly erroneous that no use could be made of them."[157]

In June 1849 Burt examined some original survey work near Higgins Lake, named for Sylvester W. Higgins, former chief topographer for Michigan's geological survey. The area had been surveyed in 1839 by another surveyor whose contract had been guaranteed by John Mullett.[158] Burt found the section lines had never been closed, and nearly half the lines and corners described in the field notes had not been marked on the ground. In his letter to Lucius Lyon, Burt concluded:

> The mystery is solved in my mind how many deputies accomplished so much work in a short time.[159]

Lucius Lyon's worst fears had been confirmed, and extensive resurveys would be required. Most of the defective work had occurred long before Lyon became Surveyor General, with one notable exception. On July 19, 1845, less than one month after he took office, Lyon contracted with his close friend Sylvester Higgins to survey in the U.P. near modern-day Munsing and Mass City. Lyon recalled, "I, in common with the whole community, had the fullest confidence [in Higgins' integrity]."[160]

In October 1849, after receiving complaints from landowners that Higgins' 1845 survey was inaccurate, Lyon instructed William Austin's son William to examine Higgins' work.[161] William examined only three of Higgins' lines and determined that Higgins' field notes were so false and fraudulent there could be no doubt as to the erroneous and fraudulent character of other portions of his work not yet examined.[162]

William Austin Burt concluded his examinations and returned to Macomb county on September 23, 1849. He was relieved that an inspection of the lines run by Alvin and Austin had confirmed their accurate work. William Austin personally checked one of their 12 townships (T25N, R11W), and reported:

> This survey was executed with a Solar Compass & the lines so far as examined bear good testimony to the utility of that instrument.[163]

After receiving Burt's report Lyon informed GLO Commissioner Butterfield that at least 150 of 170 townships in the area north of the Grand and Saginaw Rivers would have to be resurveyed at a cost exceeding $50,000. Concerning the work of Sylvester Higgins, Lyon observed, "all that he has done is worse than useless."[164]

In November 1849 the U.S. District Attorney initiated lawsuits against the surveyors and their sureties connected with the fraudulently performed contracts.[165] As a practical matter, collecting on the bonds was unlikely. In the past the government simply didn't sue, because frontier juries refused to convict men for perjury.[166] Now, however, large-scale fraud was involved, and tough action was required.

35.   William Austin Burt - face and signature *(Silas Farmer, THE HISTORY OF DETROIT AND MICHIGAN (1889), p. 1180)*

Lucius Lyon soon discovered that many of the deputies who had committed fraud were now poor. Sylvester Higgins had abruptly left for California in search of gold, but Lyon reported that the bonds on his contract were sufficient to avoid any loss by the U.S. Government. Action was then taken to secure the property of Higgins and the other surveyors and their sureties involved in the lawsuits.[167]

For John Mullett the pending suits against him as guarantor of several contracts were a slap in the face for a distinguished career of honest hard work. The biggest blow, however, was the knowledge, revealed by Burt's examination, that Mullett's son James had committed fraud.

The prosecutions against the surverors were bitter pills to swallow. Many had already endured considerable hardship, including meager pay and often unbearable working conditions. William Austin had seen men with iron determination weep like babies after several months in the wilderness.[168] Perhaps it is surprising there was so little dishonesty among surveyors presented with these circumstances.

William Austin placed part of the blame for defective or fraudulent work on the basic system of conducting surveys. In November 1849 he informed Lucius Lyon of his decision to meet with officials of the General Land Office:

I have concluded to leave home for Washington City in about four weeks & Wells Burt will go with me, if he can get ready. If you have anything to suggest to me please write. I shall make the subject of varifying Surveys done by the Deputies a part of my study that I may have something of importance to offer to the Commissioner that may prevent in [the] future fraudulent & defective Surveys.[169]

While in Washington Burt also planned to investigate the options concerning his solar compass patent, due to expire in February 1850.[170] He had personally borne the expense of perfecting his invention during the previous 14 years, with little return. The government had been the sole beneficiary of accurate work that could not have been duplicated with any other available instrument. It was about time that Burt was compensated for his efforts. Unquestionably, this would be his most crucial trip to Washington.

Washington D.C. in December 1849

Washington was in turmoil when William Austin and Wells arrived in early December 1849. Congress had convened eight days earlier, and Capitol Hill was gearing up for one of the greatest debates in Senate history over the issue of slavery.[171] The topic heated up when new territory was acquired in the war with Mexico. The South demanded that Congress divide the area into slave states, while the Northern states were adamant that slavery should be prohibited.

The House of Representatives was about to elect a new Speaker, and a new Surveyor General of Michigan was to be appointed by the President. Perhaps it is only a coincidence that less than one month after Lucius Lyon revealed to Commissioner Butterfield the full magnitude of the fraudulent and defective surveys in Michigan, Lyon's replacement was being sought. Traditionally, however, the office was filled by a political appointee, and it is not surprising that the Whig administration would replace Lyon, a staunch Democrat.

William Austin found that the party out of office is out of touch with matters such as political appointments. With regard to the selection of a new Surveyor General of Michigan, he informed Lucius Lyon:

> The councils of the nominating power are still locked up in secrecy, and no democrat at least is permitted to learn anything from them about it.[172]

It was not announced until two weeks later that Charles Noble, of Detroit, had been appointed.

During his first four days in Washington Burt conferred with many government officials, including GLO Commissioner Butterfield and Clerk of Surveys John Wilson.[173] William Austin then spent considerable time drafting his proposal for improving the public land surveys. He prepared no fewer than nine separate drafts of a letter to John Wilson, indicating the extreme importance he placed in this task. The interests of the U.S. Government, the contract surveyors, and especially the land purchasers were at stake. As he pointed out in one of his drafts to Wilson:

> The occupants of land in [areas of defective surveys] cannot satisfactorily to all concerned adjust the boundaries to their lands. Thus neighbors, if not neighborhoods, are brought to conflict which often ends in litigation.[174]

His plan called for subdividing townships into sections with the use of the solar compass. William Austin felt it was impossible to accurately subdivide land in the manner required by the surveyor's current instructions. Burt's method would ensure that sections would "approach more nearly their lawful size of one mile square."

William Austin was critical of inexperienced surveyors general who often appointed deputies ill equipped to do the work properly. These men were generally recommended by powerful political forces which vouched for the individual's qualifications.[175] As a remedy Burt suggested:

> For the best interest of the public surveys, all new deputy surveyors, and some of the old ones too, should pass an examination before three or more experienced deputy surveyors and obtain their certificate of qualifications . . . before they could be employed.[176]

He added that each chain man and marker hired by the deputy should provide written certification attesting to the individual's high moral character and faithfulness. In effect, William Austin Burt recommended some of the basic standards for the surveying profession.

Burt also suggested the routine use of an inspector of surveys. An experienced and trustworthy deputy would be appointed by the Surveyor General to examine the work of each surveyor during or shortly after the performance of each contract. Furnished with a separate map of the surveying district, the inspector would "take notes of corners and other objects as the Surveyor General may require."[177]

The survey inspector's notes would then be compared with those of the contracting deputy to ensure correctness. As a check on the inspector, the Surveyor General would require specific information together with a written diary of his activities in the field. William Austin was confident that this plan would help reduce fraud and expensive re-surveys.

Burt's proposal for an inspector of surveys was adopted. An independent deputy, unknown to the surveyor whose work was to be examined, was hired by the government to ensure more faithful and accurate work. The occasional practice of withholding part of the deputy's compensation to cover expenses of an examination was discontinued.[178] The best safeguard against fraud was the surveyor's understanding that his work would be "rigidly examined."

While the plan worked fairly well, the survey inspectors were not always compassionate. One surveyor, while running lines in Minnesota, was passing through an area when his Burt Solar Compass became entangled in dense brush. The declination and latitude arcs on the instrument were slightly bent. The maladjustment was not noticed until after the inspector, following closely behind, found the erroneous work and had the surveyor replaced.[179]

It was apparent to Burt that his solar compass would play an important role in reducing error in future surveys, particularly those in the mineral lands of the West. Commissioner Butterfield had already received a request from C. E. Booth, Surveyor General of Iowa, to replace the standard magnetic compass with a solar compass. On December 18, 1849, only days after his meeting with Burt, Butterfield informed Booth:

> Your request is approved in purchasing the solar compass, the necessity of which is manifest from the statement in your letter. As, however, one compass is all that is required by your office, it is suggested that the plain standard compass, turned over to you by your predecessor, should be sold.[180]

When he arrived in Washington, William Austin pondered the best course of action to take regarding his solar compass patent. Land office officials Butterfield and Wilson regarded the instrument "of great importance to the Government" and advised Burt to apply to Congress for compensation instead of renewing his patent.[181] They promised him a favorable report from their office and expressed their belief that reasonable compensation would be granted.

For several days Burt considered his options. The role of Congress in matters involving patents was spelled out in the U.S. Constitution:

> The Congress shall have power . . . to promote the progress of science and useful arts, by securing for limited times to authors and inventors the exclusive right to their respective writings and discoveries.[182]

The patent rights granted to Burt 14 years earlier were about to expire. He could apply for an extension of seven years, but all applications for patent extension or renewal had to be initiated at least 60 days prior to the expiration of the patent. William Austin, therefore, had until Christmas eve 1849 to exercise the option. The power to extend patents was given to the U.S. Commissioner of Patents in 1848. Congress retained the power to grant an individual patent by special Act, but it was rarely used.[183]

36.   U.S.Senator Alpheus Felch (*Henry M. Utley and Byron M. Cutcheon,*
MICHIGAN AS A PROVINCE, TERRITORY, AND STATE, THE
TWENTY-SIXTH MEMBER OF THE FEDERAL UNION, 1906)

U.S. Senator George W. Jones from Iowa recalled Burt's visit to Washington and wrote:

I was well acquainted with the Commissioners of the General Land Office, Messrs. John Wilson and Justin Butterfield, and know that they used their influence with Mr. Burt, as I and my colleague, Mr. Cass, and indeed, nearly every other Senator in Congress from the Northwest did, to induce Mr. Burt not to renew his patent for the compass, but to rely upon the Government to pass a law to compensate him for the use by the Government of his invention.[184]

On December 18th, one week before the deadline to renew or extend his patent, William Austin, still undecided, wrote Phebe:

I think I shall have no difficulty in obtaining a new patent for my Solar Compass with the improvements, or selling out to Congress. The latter I should prefer, if I can do so.

86

37.   U.S. Senator Lewis Cass (*Henry M. Utley and Byron M. Cutcheon,*
MICHIGAN AS A PROVINCE, TERRITORY, AND STATE, THE
TWENTY-SIXTH MEMBER OF THE FEDERAL UNION, 1906)

Burt appeared more interested in the future of the public land surveys than the potential income from royalty fees. At the time, however, he had little barganing power. His invention had been limited to the U.S. land surveys, and if the government restricted its use, the solar compass would not have a market. With a government lawsuit filed against him as surety in the Nicholson contract, and his friends in trouble from his incriminating examinations, William Austin's position was weak.

There was little choice but to accept the advice of those who counselled him. On New Years eve 1849 Burt revealed his decision in a letter:

> I wish to say to you that [the] government is about to order the use of the Solar Compass in all Surveys of the U.S. Lands. It is recommended to me here by men in high authority to apply to Congress for remuneration for its past & prospective use in the public surveys. I propose not to renew my patent or secure the right to the important improvements which I have made in this instrument in order that they may be sold to surveyors at the lowest possible price, that they may be brought more spedily into general use.[185]

Surprisingly, Burt apparently did not consult with Lucius Lyon, co-owner of the solar compass patent, before he made a final decision. For a $15 fee and an application filed before the deadline, Burt and Lyon might have earned substantial royalties by a patent renewal or extension. Instead, Burt broke the news to Lyon in a letter written on New Year's day 1850:

> I find here that the time has expired for getting an extension of my patent or securing the right to the many improvements which I have mad[e]. I am counceled here to petition Congress for relief in this matter in which I have spent so much time for the public benefit. . . . It would afford me pleasure to return to you the funds you let me have on account of the Solar Compass and if success is had in my present effort I shall be able to do so.[186]

Burt's action confirms the confidence he placed in men like Senators Cass and Felch. Although Cass had been narrowly defeated in his 1848 bid for the Presidency, he was re-elected to the Senate and remained a powerful political influence. Senator Felch was President Taylor's strongest anti-slavery spokesmen and Chairman of the Committee on Public Lands.[187]

On January 4, 1850, William Austin sent his petition for compensation to Senator Felch. In it he explained that he had spent 14 years and considerable expense improving the instrument. He had also waived most of the royalty fees so more surveyors could afford to purchase the solar compass. Documents were included to support his claim that:

> without the use of this instrument the mineral lands in Michigan, Wisconsin, Iowa, and Arkansas could not have been surveyed for the Government for less than double, and probably three or four fold, the amount they actually cost; as in most of those mineral regions the ordinary compass was of no use. . . .[188]

Now the matter was in the hands of Congress.

PART THREE

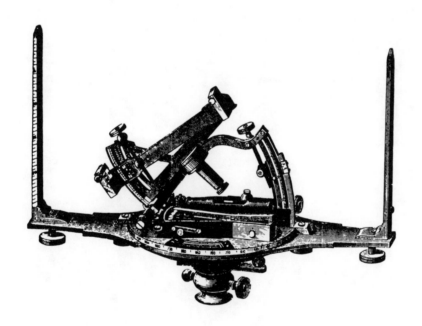

Chapter XIV

JUSTICE FOR ALL

*Why should there not be a patient confidence
in the ultimate justice of the people? Is there
any better or equal hope in the world?* [1]

- Abraham Lincoln (1809 - 1865)

Precedent

Burt's decision to petition Congress in lieu of renewing his solar compass patent came soon after the House concluded its three-week struggle to elect a Speaker. After 63 ballots the Democratic candidate, Howell Cobb of Georgia, was chosen by a two-vote plurality over Whig Robert C. Winthrop of Massachusetts. It was the first time in history the House abandoned the principle of majority rule. Democrats and Whigs were nearly equal in strength, leaving the balance of power in the hands of 13 Free Soil Republicans.[2] In retrospect, however, Speaker Cobb's pro-slavery position was a triumph for the South, and his selection served to aggravate the struggles over this issue during the 1850's.[3]

The Democrats had solid control in the Senate during the 31st Congress, and from that chamber Burt expected strong support for his claim. For ultimate success, however, a bill would have to follow a lengthy course.

First, Senator Felch must prepare a bill from Burt's petition. It would then be introduced to the full committee of the Senate and then sent to a specialized subcommittee for study, hearings, revision, and approval.

If all went well, the bill would be returned to the full committee for more discussion and revision. If approved, it would be sent to the House and the process repeated. If the bill was approved by both chambers, it would then be enrolled on parchment and sent to the President for final approval. If it failed to receive the approval of both chambers before the final session of Congress adjourned, a new bill would have to be prepared.

In reality, as Burt's advisors must have known, most bills involving private claims never become law. It is also unlikely that a successful precedent had been established. During the First U.S. Congress, in 1789, Surveyor-Cartographer John Churchman III of Nottingham, petitioned Congress for funds to make a trip to Baffin Bay, where he was convinced the magnetic north pole was located. He had invented a new method to explain the principles of magnetism, but despite James Madison's support for the expedition his request was denied for a lack of funds.[4]

In 1838 scientist William C. Poole was equally unsuccessful with his petition to Congress. Like Burt, Poole had found the magnetic compass a "very imperfect instrument."[5] Consequently he invented an improvement in the magnetic dipping needle and an accompanying instrument for finding the latitude, longitude, and the variation of the compass. He urged that Congress approve funds for several copies of a quarterly report of magnetic variations which he believed would be useful to "all mariners, surveyors, and engineers in our country". No action was ever taken on his request.

Another inventor, Moses Smith of New York, had attracted the interest of the Navy Department with an improvement in the the mariner's and surveyor's compass needle. It was much heavier than conventional magnetic needles and it included special sliding caps. Smith claimed the needle was much more stable when subjected to violent concussions, such as the firing of a ship's cannon.[6] He requested an unspecified amount of money from Congress for any future benefits that may be derived from his improvement. Smith chose to petition Congress rather than "submit his right to the uncertainties and abuses of the Patent laws." Several of his improved compasses were provided to the Navy Department and tested on naval and commercial ships during the following year. In 1833 Smith's needles were judged superior to others in use, but it does not appear that he was ever compensated.[7]

William Austin left Washington on January 10, 1850, encouraged by his advisors that his petition would be successful. The following month, from his home in Mt. Vernon, he sent the following message to John Wilson at the General Land Office:

> In regard to my petition to Congress for remuneration for time and expenses in furnishing the Solar Compass for the survey of the Public Lands, I leave it for you and Senator Felch to dispose of as you and he may see fit for I cannot be in Washington to attend to it.[8]

As William Austin's son John later acknowledged, "He chose not to hold the staff in his own hand".[9] If he had applied to the Patent Office he might have succeeded. Of the 17 applications to extend patents due to expire in 1850, only 5 were rejected. There were 27 patents re-issued during the year, three for additional improvements.[10] Now all he could do was wait.

Burt's Astronomical Compass

On his way home from Washington in January 1850, William Austin visited William J. Young in Philadelphia and supervised construction of Burt's new invention. The idea was first presented to Young in the winter of 1847.[11] Burt called the instrument an "Astronomical Compass." It incorporated the principles of the solar compass with additional improvements, including a complete circle for determining the approximate longitude. It could also be operated during the day or at night.[12]

Burt predicted his device would become "a most useful instrument in the geological surveys and other exploring expeditions in our new country."[13] In a letter to John Wilson, he said:

> I hope the attempt to make an important improvement will be successful for it will be quite an expensive one.[14]

By March 1850 the Astronomical Compass was nearly completed. Four of William J. Young's workers had labored on the instrument and Young personally marked it with precise graduations. To Burt he enthusiastically wrote:

> I have had it pretty much all together today, and it was a very fine appearance. Both telescopes are done and fixed, and its numerous tangents, the circle of arc, makes it look imposing. I feel sure you will be pleased with it.[15]

In May 1850 William Young finally sent the new compass to Burt with an apology for the lengthy delay. According to Young, the workman who finished the main component "only works to live, and does not live to work, and during a goodly part of the time he lived up to his doctrine."[16] Young charged Burt $250 for the Astronomical Compass, although he estimated it took at least $300 in labor and materials to make it.

38.   J. Bailey advertisement *courtesy of Marquette County Historical Society*

Burt's new invention was field tested in the Upper Peninsula by geologist Charles Whittlesey, who called it "Burt's Solar Compass for longitude."[17] In December 1850, Burt informed Young that the arcs of the instrument were in serious index error.[18] Five months later, from London, Burt notified his wife that, "I had no objection to the sale of my astronomicle compass, but shall not order another one untill I returne."[19]

Detroit instrument maker John Aylesworth Bailey advertised in 1852 and 1853 that he could make either Burt's Solar Compass or Burt's Astronomical Compass.[20] It is not known, however, if the latter was ever put into practical use. Conceivably this modified solar compass was the precursor of the universal instrument developed by Burt in 1856.

Rectifying a Misconception

In March 1850 William Austin sent a letter to the editor of Michigan Farmer, which later appeared on the front page of the April 1850 issue.[21] Burt sought to correct the widely held misconception that the land between Saginaw and Traverse City was worthless for cultivation.

The discrepancy first became apparent to William Austin during his examination of the Higgins Lake area the previous June. When the GLO Commissioner dropped plans for new surveys in the area, Burt told Lucius Lyon:

Perhaps [the Commissioner] thinks this country is not worth surveying, but when I return home, I shall try to bring that question to a proper test.[22]

93

Burt was confident that much of the soil he had inspected could be quite productive with proper cultivation. He described the area near Grand Traverse as "good farming country," and "for beauty of surface and fertility of soil, swamps excepted, it is not much, if any, excelled by any portion of our state."

In 1852, Surveyor General Charles Noble confirmed Burt's observations and explained why the area had previously been neglected:

> . . . much of the country heretofore represented in the original surveys as indifferent second and third rate land, and swamp, and lake, is proven, by the resurveys, to be among the choicest land in the lower peninsula of Michigan.[23]

The surveyor had drawn lakes on the map where none existed, postponing the sale and settlement of this prime land.

Burt's favorable assessment of the soil proved to be accurate. Orchards of cherry trees planted later near the village of Traverse City soon flourished, and today it is claimed that nearly one-third of the world's cherry crop is produced from this region.[24]

Pending Litigation

By mid-1850, the consequences of Burt's 1849 examinations were still being felt as the government's lawsuits against several surveyors and their sureties continued. In July William Austin prepared to make good on his security bond for Henry Nicholson's defective work in 1837. While in Washington in December 1849, Burt had received permission from GLO Chief Clerk John Wilson to resurvey Nicholson's district, and later Michigan's Surveyor General explained the decision to the GLO Commissioner:

> Judge Burt in this matter seems to have been prompted by a desire to serve the public interest. It was at his suggestion that he entered upon the work. He does not, however, claim any compensation for this service, but submits the whole matter for the decision of the Department.[25]

During the following months Burt completely resurveyed five townships and corrected five others in Nicholson's district, at a personal cost of $3,000. Although the Surveyor General recommended some compensation for this work it was never paid.[26] This may explain why William Austin did not accept any more major surveying contracts on the public land surveys. His last survey work, completed in October 1853, comprised a small area of private claims near Milk River Point in Lake St. Clair, Michigan.

Lucius Lyon was much less fortunate than Burt. During 1850, after he completed a resurvey in the Lower Peninsula, Lyon discovered that the government had seized and sold a large parcel of his property near the Mississippi River to satisfy a judgment against him as surety for another surveyor's contract. Lyon had been unaware of the sale, although it had occurred several years earlier. He never recovered from this financial blow, and on September 24, 1851, he died after a brief illness.

John Mullett was anxious to remove the stigma of lawsuits against him as surety for other surveyors. He asked for the same opportunity granted to Burt, to re-survey the areas of defective work at his own expense. In March 1850, Surveyor General Charles Noble asked GLO Commissioner Butterfield to approve the request of Mullett, whom he considered:

> one of the most competent and faithful men that has ever been sent into the field as a deputy. No suspicion has ever, I believe, been attached to any work performed by him in the survey of the public lands.

Four months later William Austin inquired about Mullett's request and was informed by Noble that Butterfield had ignored it. In November 1850 the lawsuit against Burt was dropped, as he had more than made good on his bond. Although a Detroit jury acquitted James H. Mullett of performing fraudulent work, his career as a government surveyor was over. Three years later another examiner found an earlier survey by James Mullett had been "most fraudulently" performed.

Eventually John Mullett cleared his record and returned to surveying. In October 1851 he was paid $5 per day plus expenses to run a preliminary survey of the Sault Ste. Marie village. Although he completed his assignment, he declined an invitation to survey in Sault Ste. Marie the following winter. Instead, he retired with 30 years of distinguished service as a U.S. deputy surveyor. It was a record unsurpassed by any of his Michigan colleagues. On January 10, 1862, nine years after he moved to Ingham County, near Lansing, John Mullett died at the age of 75.

There is no evidence that the friendship between William A. Burt and John Mullett ever deteriorated despite the difficulties incurred by the Mulletts following Burt's 1849 examinations. In 1888, however, one of John Mullett's sons would initiate an act of retribution against the Burt family.

39.   Witness tree marked by
       William Burt - June 17, 1850 *(author file)*   William Burt *courtesy John H. Clarke*

The Witness Tree

The year 1850 was the last in which William Austin and his four remaining sons all worked on the U.S. public land surveys. Wells and John subdivided in the Upper Peninsula while their father corrected the errors in Nicholson's earlier work. John was the only surveyor to work in the mineral region during the year and provided valuable assistance to geologists Foster and Whitney. Austin and William subdivided townships in the Lower Peninsula.

During his survey, son William marked a red pine tree near Vanderbilt, in Otsego County. Today it stands as "perhaps the most famous witness tree in Michigan."[37] The tree was blazed by William on June 17, 1850, to identify the quarter section corner, and it measured only eight inches in diameter. Today it has grown to a majestic height of over 100 feet, with a diameter of over 30 inches. Maintained by Michigan's Department of Natural Resources, it stands as a living monument to William Burt.

Solar Compass and the Instructions

In January 1849 William Austin proposed a new method for subdividing the Northern Peninsula with the use of solar compass.[38] In his letter to Surveyor General Lucius Lyon, Burt wrote:

> By your request I have been studying and writing an outline of a system of subdividing townships into sections on the northern peninsula of Michigan. I have aimed at this, on the most practical and scientific principles I know of and hope to attain the most safe, cheap, and best for the public interest; and well adapted to the use of the solar compass.[39]

Burt's plan, dated January 17, 1849, was later incorporated in the 1850 "General Instructions To His Deputies; by the Surveyor General of the United States, for the States of Ohio, Indiana, and Michigan." Included for the first time was the following requirement:

> All surveys of every description, where the magnetic variation is not uniform, must be made with Burt's Improved Solar Compass, or some other equally good instrument. . . . When the sun shines, any survey may be accurately and expeditiously made without the use of the magnetic needle.

Reference to the "improved" model undoubtedly precluded the use of earlier less efficient versions. After his patent expired Burt made no further improvements to his solar compass, although individual manufacturers added their own modifications over the years.[40]

To explain changes in the variation of the compass, a table of observations compiled by William A. Burt in 1839 was included with the 1850 General Instructions. In addition, a "Specimen of Field Notes" contained a sample plat map depicting three adjacent land claims held by Lucius Lyon, William A. Burt, and John Mullett. The designated location of the fictional township is near the center of Lake Superior.

The Iowa-Missouri Boundary

The use of the solar compass was soon expanded to other states. The Surveyor General for Iowa, Caleb H. Booth, considered Burt's Solar Compass a great advancement in accurately surveying the public lands.[41] On May 24, 1850 he sent his deputy William Dewey, from Iowa, to join surveyor R. Walker, of Missouri, to survey the boundary between the two states.

The western portion of the line, from the old western boundary of Missouri to the Missouri River, was required to be a "parallel of latitude" and Burt's Solar Compass was used. William Austin defined parallels of latitude as "curved lines that increase in curvature from the equator to the poles, and cross all meridians at right angles." At the end of each mile of line run, a northern offset of 6.855 inches was made to correct for convergence.

The variation of the needle, as shown by the solar compass, was noted every quarter of a mile. The original line was surveyed by John C. Sullivan who used only one variation of the compass. Dewey and Walker found that his line rarely ran in the same direction for any two consecutive miles.[42]

The Iowa-Missouri boundary line thus became the second state line run with Burt's solar compass.

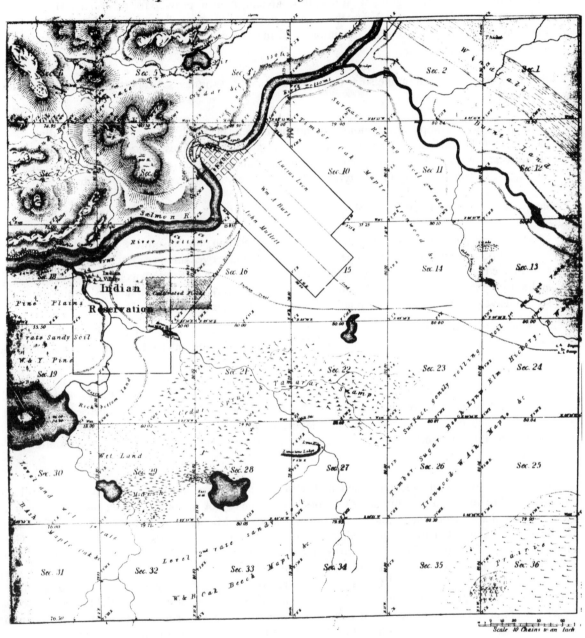

40. Fictitious township - General Instructions of 1850 (Ohio, Indiana, Michigan)

Burt's Petition

By mid-March 1850, Senator Felch had received several letters supporting Burt's petition. One was from Charles Whittlesey, U.S. geologist for the Lake Superior region, who regarded the solar compass as:

> an ingenious instrument, constructed on scientific knowledge by a mathematical mind. It is now used in the surveys of Michigan and Wisconsin, because no other compass will make lines close in the mineral region, but it would be advantageous to provide for its use elsewhere, and in all the public surveys. From the small number now in use I am satisfied that Judge Burt has received little from his invention, and the Government has already derived great benefit.[43]

Whittlesey later commented on the solar compass in Foster & Whitney's "Report of the Lake Superior Land District":

> Without the Solar Compass the region could not have been surveyed, except at a cost exceeding the value of the land. . . . The correction, base, and meridian lines, since the introduction of the solar compass, have been generally surveyed with much care, particularly the principal meridians; so that they may be regarded as following very nearly parallels of longitude.[44]

Geologist-inventor John Locke, commenting on the inaccurate magnetic compass needle, wrote:

> I have seen it point, not only at right angles to the meridian, but absolutely inverted by the north end pointing to the south; and yet Burt's township line over this very tract was correctly run. I have been enabled to fully appreciate the value of such an instrument as he has invented, and am prepared to say that by the improvement of the work of our land surveys, which he so extensively introduced, he has rendered as essential national service.[45]

GLO Commissioner Butterfield forwarded John Locke's letter to Senator Felch and included his own testimonial. Referring to the use of Burt's Solar Compass in the mineral lands of the north, Butterfield wrote:

> . . . the work was not only well done, but much more economically than could be done by any other instrument now known. . . . The Government, then, really has derived all the benefit which has flowed from the time, labor, study, and expense incurred by Mr. Burt in perfecting this instrument, and it is but a simple act of justice to allow him a fair compensation thereof.

Commissioner Butterfield recommended that a rider be added to the general appropriation bill "to compensate Burt for the use of his Solar Compass in the survey of the public lands, thousand dollars." He estimated that "before the whole of the mineral lands in Michigan, Wisconsin, Iowa, Minnesota, California, and Deseret shall be surveyed, hundreds of thousands of dollars will be saved to the Government. . . ."[46] The issue of Burt's compensation, however, was not included in a general appropriation bill; in fact, no action was taken on Burt's petition until 1852.

Nevertheless, the General Land Office was already taking steps to bring Burt's Solar Compass into general use in the public land surveys. In November 1850 Commissioner Butterfield reported to the Interior Secretary that he intended to break with tradition by purchasing surveying instruments, including the solar compass.[47] They were to be sold to the surveyors at cost, with an option to repurchase.

In September 1850 the office of Surveyor General for Oregon was created by an Act of Congress. Two months later John Preston of Chicago was appointed to head its operation. One of Preston's first priorities was to order several solar compasses from William Young. He offered to pay a premium price if Young could deliver them in time for the Oregon surveys, but Young could not accommodate him.[48] Young had as many as five men finishing work on six solar compasses, but there were also many additional commitments to keep. The problem would have to be resolved.

Preparations for the Great Exposition

It was late 1850 when William Austin decided to take part in the first international exposition to be held in London's new Crystal Palace the following year. He informed William Young that he would enter his solar compass among the American exhibits, and he ordered one for this purpose.

Burt undoubtedly considered the world's fair an opportunity to expand the use of his solar compass to other countries. He received little encouragement from Young, however, who told him:

> The solar compass I don't think will ever be used much in Europe, 1st because it's use will not be required there as much as here, and 2nd because the kind of instruments they have been accustomed to answer all their purposes, and you could get but few to change. In many colonies of Great Britain, it would, as here, be eminently useful, but that would have to be a government affair. If you wished to urge, I suppose the government would give you a patent, but patents are costly affairs.[49]

Despite Young's lack of enthusiasm, William Austin was confident his first trip to Europe would be a memorable experience.

Chapter XV

NEW FRONTIERS

Bring me men to match my mountains;
Bring me men to match my plains, -
Men with empires in their purpose,
And new eras in their brains. [50]

- Sam Walter Foss (1858 - 1911)

41.   The Crystal Palace *(Tallis's HISTORY AND DESCRIPTION OF THE CRYSTAL PALACE, n.d., London & N.Y.)*

The Great Exposition of 1851

It promised to be a glorious show, the first international gathering of its kind. Prince Albert, husband of Queen Victoria, received approval of the plan in 1849 and invited all the civilized nations to come to London to display their raw materials, industrial products, and artistic wares.

The objective of the Great Exposition was to show how each nation had developed, and to provide a new starting point for further achievements. The friendly rivalry among nations would culminate with the awarding of bronze medals for the items judged best in each of several categories.

At first, U.S. reaction to the event was lukewarm and skeptical. Writers for the Whig Review and other newspapers feared that the Americans would make a poor showing and attacked the exhibition. Soon, however, sentiment warmed to the idea. A New York Herald editorial proclaimed:

> We are anxious to know what success the mechanics of the U.S. will meet at this fair. We are convinced they will take their due share of the premiums and will give the world a better appreciation of what the republic is. . . . The more we are tested, the more we triumph.[51]

The exhibition was not scheduled to open until May 1st, but William Austin left Detroit early, arriving in New York City on April 13th. Phebe accompanied him as far as Buffalo so she could visit their friends and relatives in Wales Center.

Burt carried with him a letter of introduction from Senator Lewis Cass to A. Laurence, U.S. Minister to England. It read:

> The bearer of this letter, Judge Burt, is a worthy man and a highly respectable citizen of this State. He visits London for the purpose of witnessing your great show, and I beg leave commend him to your attention, should circumstances render it necessary for him to call upon you. I have a sincere regard for him.[52]

At noon on April 16, 1851, William Austin departed New York on the U.S. Steamship Baltic for Liverpool, England. Among the passengers was Horace Greeley, Editor of the New York Tribune, and known today for his advice to a generation of American Youth, "Go West young man."[53] Greeley had just written one of his many stinging antislavery editorials, and he would serve as chairman of one of the juries awarding medals at the fair.

The Atlantic crossing took 13 days, and Burt quipped, "The Ocean did not fail to roll its majestic billows, continually."[54] William Austin was unaffected by the rough seas, but Greeley was seasick most of the voyage. During the 240 mile train ride from Liverpool to London, Burt noted, "This ancient country, all cultivated like a garden, [is] unlike anything that I have seen in America."

The Crystal Palace, erected for the Great Exhibition near Hyde Park, was the most imposing sight of all. Designed by Sir Joseph Paxton, head gardener to the Duke of Devonshire, it was like a gigantic greenhouse of glass and iron. It was the first prefabricated iron-framed building, and its walls and ceiling contained nearly 300,000 panes of hand-blown glass. Burt's letters fail to reveal his thoughts of the "glass palace," but nearly everyone was impressed. English mathematician Charles Babbage, a pioneer of the modern computer, wrote:

> There will be found within that crystal envelope, few whose manufacture can claim a higher show of our admiration than that palace itself, which shelters these splendid results of advanced civilization.[56]

Somehow Burt managed to obtain one of the 25,000 tickets distributed for the opening ceremonies on May 1st. He witnessed the highlight of the day, a procession by Queen Victoria, Prince Albert, the Royal entourage, and a multitude of dignitaries from around the world. When the Queen officially declared the exhibition open, the halls erupted with organ music featuring the Hallelujah Chorus.

Like nearly everyone present Burt was emotionally affected. He wrote, "This grand procession consisted of nearly every nation . . . the like [of which] was never seen before." Queen Victoria called the occasion "more touching" than even her own coronation.[57]

William Austin was simply glad to be there, impressed by everything in sight. In a letter to Phebe he attempted to share the occasion:

> I cannot give you anything like an intel[l]igent view of it. It exceeds by far anything and everything that I had imagined. There is here about everything to be seen in nature and art, that this world produces.[58]

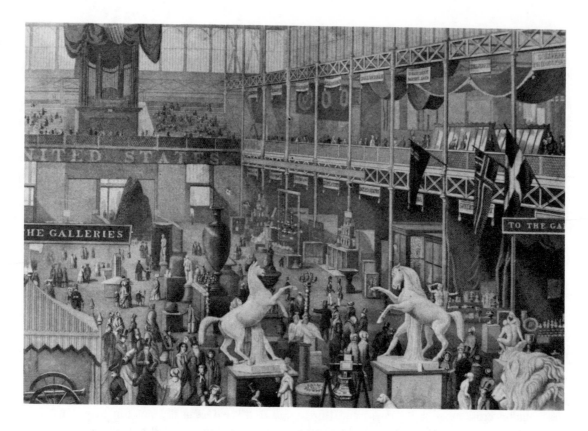

42.   American exhibits at the Great Exhibition *(Patrick Howarth, THE YEAR IS 1851, London: 1951)*

Later Burt wrote:

> I have attended the great Exhibition every weekday except one since I have been here in London. The world has never seen the like before, nor will it see it again soon, for the world cannot afford an exhibition of this kind often. . . . Best of all the nations of this world are here in perfect peace and show much goodwill and friendship towards one another.[59]

William Austin was not optimistic about America's chances for winning awards at the Great Exhibition:

> I do not expect much from [the judges] favourable to Americans, for the U.S. have not sent so much to the exhibition as most all nations expected of them and what we have is of the plain useful kind and does not make a show like the ornamental.[60]

Even Queen Victoria found the American machinery "inventive, but not entertaining."[61] Of the 4,000 exhibits, only 599 were from America. Burt's Solar Compass, displayed unobtrusively in the "Astronomical Instrument" section, was quite a contrast to the massive, noisy machinery that was displayed nearby.

During June and July, William Austin, accompanied by an interpreter, found time for sightseeing in Europe. Wearing his wig, as he generally did when he dressed up, he toured France, particularly Paris and Versailles, Northern England, and Scotland.

In Scotland Burt was welcomed by Hugh Miller, an eminent Scottish geologist and author whose writing impressed him. In England he met with Sir John Herschel, distinguished chemist and president of the Royal Astronomical Society. Herschel had high praise for the solar compass and told Burt:

> I have long understood the elements of your instrument, but could not see how they could be carried out mechanically. It has fallen to your lot, Sir, to not only conceive the necessary astronomical elements, but also to carry them into practical effect, mechanically.[62]

It was with these men that William Austin felt most comfortable. He told Phebe:

> I have had an op[p]ortunity to associate with Lords and Noblemen and to be in the presence of Queen Victoria & Prince Albert and other Princes, for whome mutch greate veneration is had in England. as if the gods were present but I could not feel any such veneration for them, for I saw and felt that they were human beings, but a worm of the dust like myself. My greater enjoyment has been with scientific men who are here at this time from every quarter of the world.[63]

William Austin's visit to the Great Exhibition was a religious experience as well as an intellectual adventure:

> God may use this great peace convention of all nations to advance his cause and save men from death. This is a strange world, or rather a strange fallen and accountable race of men inhabit it, and I often think what can be done to better those conditions. The answer returns to me, the Gospel of the grace of God alone can do it.[64]

Burt remained in Europe over two months, hoping to be on hand when the medal winners were announced. When he learned the awards would not be finalized until October, however, he arranged to return to America. He notified Phebe that on the way home he would spend two days in Philadelphia with William Young to complete some new work that required his personal supervision.[65]

On October 15, 1851, the Great Exhibition closed. Queen Victoria praised the occasion as the "happiest, proudest day in my life."[66] The first World's Fair had been a pinnacle in the career of Prince Albert. With over six million people attending it easily returned a profit to its financial backers. Even Horace Greeley was impressed, calling the exhibition, "The first satisfactory model of a Peace Congress."[67]

About 25% of America's exhibits received awards, including several "honorable mentions." Among the five Americans receiving "Council" medals were Cyrus McCormick for his reaper and Charles Goodyear for rubber exhibits.

Recipients of the "Prize" medal included William Austin for Burt's Solar Compass and Samuel Colt for the Colt revolver. A certificate, dated October 15, 1851, accompanied Burt's bronze Prize medal. Signed by Prince Albert, it read:

> I hereby certify that Her majesty's Commissioners, upon the award of the jurors, have presented a prize medal to William A. Burt, for a Solar Compass and surveying instrument, shown at the exhibition.[68]

The award was a great honor for William Austin, who accepted it with pride. Ironically, while winning the Prize medal, he was still having to prove the value of his solar compass to the U.S. Congress. In his final letter from Europe Burt wrote, "My solar compass I think is more appreciated here than at home."[69]

43.   A U.S. Deputy Surveyor operatiing a Burt solar compass near Progreso, New Mexico
        courtesy James F. Davidson, Jr. AUTHOR'S NOTE: U.S. Deputy Surveyor Jay
        Turley was selected as a prototype land surveyor by GLO to pose for this picture that
        was displayed as part of the GLO exhibit at St. Loius Fair in 1903-04 and the Portland
        Fair in 1905. The original photograph was reproduced in water colors.

When William Austin returned home in August 1851, Congress had not yet acted on his petition for compensation. Government surveyors, however, were using the solar compass to open up the West for settlement. The 1851 Manual of Field Operations for government surveyors in Oregon required all township lines to be run with "Burt's improved solar compass, or an instrument of equal utility." When the needle could be relied on, the magnetic compass was allowed for subdividing and meandering. The 1851 Manual was immediately adopted for use in California, Minnesota, Kansas, Nebraska, and New Mexico.[70]

When Surveyor General Preston arrived in Oregon City in May 1851 he had four solar compasses with him.[71] Manufactured by William Young, they had been relinquished by the Surveyor General of Michigan to avoid any further delay.[72] Commissioner Butterfield had instructed Preston to "cause the work to be executed upon terms as favorable to the Government as is consistent with accuracy."[73] The solar compass would play an important role in this effort.

William Austin declined an invitation to go to Oregon because of his pending trip to Europe.[74] Instead, he recommended William Ives, an experienced surveyor, skilled in the use of the solar compass. Ives was awarded the important contract to survey the Williamette Meridian, north from the initial point, located near Portland, and the baseline east to the Cascade mountains and west to the coast range.

Webster Kimball was an Oregon surveyor who owned his own solar compass by the time he was appointed a U.S. deputy surveyor. Although the government was paying $190 for solar compasses, Mr. Webster had purchased his from another surveyor at the premium price of $350.[75]

In October 1851 Surveyor General Preston reported to Commissioner Butterfield the results achieved with Burt's Solar Compass:

> There is considerable local attraction found to exist throughout the whole country, so much that the magnetic needle cannot be depended upon in making the surveys. Burt's improved solar compass has been used on all the lines that have and are being surveyed, and found to be an admirable instrument, in fact the only one that can be used to advantage in the surveys of this coast.[76]

The solar compass was one of the instruments used during the the operations to establish the longitude of Salem, Oregon. The Surveyor General directed that great care should be taken in making this determination by astronomical observation and by measuring a line to the coast. The result would then be compared with those obtained by the U.S. Coast Survey. From well determined points, over 50 bearings were taken with the solar compass to the peaks of Mt. Hood and Mt. Jefferson. It was reported that "the result was highly satisfactory, as the compass work fitted admirably . . ."[77]

When Samuel D. King was appointed Surveyor General of California on March 24, 1851, Commissioner Butterfield planned to provide him with three solar compasses, borrowed from the Michigan Surveyor General's office.[78] Unfortunately, there were none to spare. Concerned, King told Butterfield:

> If we expect work to be well done even by the best men, they must use the best instruments in their field operations.[79]

King then requisitioned several instruments, including three improved solar compasses manufactured by Richard Patten of Baltimore, who began making solar compasses in the early 1840's. Patten charged the GLO $195 for each instrument, which was then sold at cost to the contract surveyors with an option to repurchase them.[80]

Eventually, most of the early surveys in the West were made with a solar compass.[81] The respect accorded to Burt's invention is evidenced in the official Seal for the Surveyor General of California, beginning in the early 1850's. A likeness of the solar compass, with an image of the sun's rays above it, appears in the center of the Seal which was embossed on the early California land documents and maps.

44.　California Surveyor General's seal depicting Burt's solar compass (1885)
courtesy Bud Uzes. Author's pic. from xerox. (From doc. approved by Surv. Gen. Richard P. Hammond, 11/20/1885.)

45.   John Burt
courtesy John H. Clarke (author collection)

Chapter XVI

JOHN BURT IN THE UPPER PENINSULA

To whom in vision clear
The aspiring heads of future things appear
Like mountain-tops whose mists have rolled away. [82]

- William Wordsworth (1770 - 1850)

John's Bold Prediction

Although William Austin Burt paid little attention to the mountain of iron he located in 1844, his son John took an early interest in developing the mineral wealth of the Upper Peninsula. In many respects John exhibited his father's characteristics more than his brothers. His obsession for accuracy and thoroughness served him well as a skilled engineer and surveyor. Physicially he was described as "tall, well-built, with a frank, pleasant face, and a very engaging manner.[83] He was also a tough-minded, aggressive visionary, determined to succeed where others had failed.

On December 3, 1835, John married Julia Calkins, and in 1844 they moved their family from Mt. Vernon to Detroit.[84] By 1850 John and Julia had three living children: Hiram, 13, Alvin, 9, and Minnie, age 3. Their daughter Mary had died of scarlet fever in 1844.

During his surveys in the mineral region of the Upper Peninsula John found several choice areas for future development. In June 1849 he became an early landowner with the purchase of 92 acres of land at the Carp River near present-day Marquette. The Jackson mining Company had built a forge near this area in 1847, and in 1848 the first charcoal iron bloom was produced.[85] By December 1849 John had purchased an additional 1,363 acres in the area, but his early presence as an Upper Peninsula pioneer has been largely ignored by historians.[86]

In 1850, at age 36, John retired from the U.S. public land surveys to concentrate his efforts in the Upper Peninsula. During the fall he pulled off an important land coup. Although he hoped to purchase a large tract of land in the iron land district, he felt the price of $5.00 per acre assigned to all mineral lands was excessive.

Consequently, John applied at the Sault Ste. Marie land office for an opinion from the U.S. Attorney General regarding the classification of the iron ore lands as mineral or agricultural. When he learned that iron land would be considered as agricultural land to be sold at $1.25 per acre, he purchased 15,000 acres which later became a part of the land holdings of the Lake Superior Iron Company.

The purchase was the first made under the new decision. A packer from a surveying party actually made the claim on John's behalf. He had built a small shack and lived on the property long enough to satisfy the requirements for residency. The claim, however, was contested by Robert J. Graveraet, a founder of Marquette. John did, in fact, give Graveraet an undivided one-half interest in the land that was then assigned to the Marquette Iron Company.[87]

46. Marquette, near the Carp River - 1851 (*Frontispiece, U.S. Cong., 32-1 (1851), SS#609, p. 15*)

John envisioned enormous profits could be made investing in the Upper Peninsula; however, when he offered a group of prominent Michigan men the opportunity to purchase a three-eights interest in his land for $50,000 they declined.[88] Undaunted, John traveled to Pittsburgh in the spring of 1851 to interest the iron barons of Pennsylvania in his plan to market the Upper Peninsula iron. Many of the early U.P. mining companies built their charcoal furnaces near the mines. John was convinced, however, that it would be more economical to export the ore to well developed markets.[89]

By July 1851 John had contracted to sell steamboat wood for a 50% profit. He estimated that his property, totaling 16,500 acres, had nearly doubled in value. One investment in a copper location north of Lake Gogebic was sold only a few months later for a tenfold return.[94]

It was during construction of his sawmill that John first met Herman B. Ely, a railroad financier from Cleveland. Like John, Ely envisioned a bright future for the Upper Peninsula. He invited John to work with him to complete a railroad from Lake Superior to the iron mines. Since a railroad was a vital element in John's plan to market the iron ore, he readily agreed to Ely's proposal.[95]

Soon Ely obtained contracts from officials of the Jackson Iron Company and the Cleveland Iron Company to build a road from Marquette to their mines. John intensified his activities in the Upper Peninsula, and on November 4, 1851, he was elected county surveyor in Marquette County's first general election.

It was a mismatch of power when John confronted Dr. Peter Shoenberger whose first Pittsburgh furnace was built when John was only three years old. Shoenberger, aged 70, was considered Philadelphia's most prominent ironmaster.[90] He told John:

We have an abundance of good ores in Pennsylvania and have no need of your Michigan ores. Besides, you will not see a ton of it in this market in your or my day.[91]

Firmly, John replied:

Mr. Shoenberger, you will have it here in five years at fartherest and beg for it.

It was a bold prediction. No one shared John's optimism, although the high quality of Michigan iron ore had already been acknowledged. In 1850 two men from Pittsburgh examined ore at the Jackson mine and told U.S. mineral agents that in their opinion it was the richest iron deposit in the United States.[92]

His resolve intact, John returned to the Upper Peninsula determined to fulfill his promise to Shoenberger. With a crew of laborers and mechanics he built a dam across the Carp River and constructed a sawmill in preparation for erecting a forge to manufacture blooms. The dam and sawmill were completed in 1852 at a cost of $6,900. Lumber cut in John's sawmill was used to construct many of the first structures in Marquette, including the first church, the first school house, the first ore dock, and the first railroad.[93]

Lobbying for the Soo

A railroad from the mines to Lake Superior would be of little value unless the ore could easily be transported by ship to the distant markets. The falls on the St. Mary's River remained a major obstacle, and the increased activity at the mines resulted in lengthy delays in Sault Ste. Marie. Ships were hauled on rollers by portage around the falls. Otherwise, the ore, packed in barrels, had to be unloaded and carried around the falls to ships waiting on the other side.

Cargo began to pile up at Sault Ste. Marie, partly due to a lack of steamers on Lake Superior. In 1851, only four steamers operated in Lake Superior, and, incredibly, two of them collided.[96] Wells Burt and his cousin Zelotas Searles were returning home from a survey, on board the Monticello, when the accident occurred. They were among the 100 passengers and crew who managed to reach shore safely. Wells lost $600 worth of equipment including two solar compasses, but his field notes were saved.[97]

The need for a canal at the Soo was now obvious, but approval of the project was far from assured. Nearly 12 years had elapsed since Michigan's efforts to build a canal at the Soo had been met by Congressional apathy and military force. This time, to avert such resistance, John and other prominent Michigan men formed a pressure group to lobby for the canal bill in Washington.

When the 32nd Congress convened, on December 1, 1851, resolutions from several state legislatures were presented in support of the bill to help Michigan construct a ship canal around the "falls of Ste. Marie." In addition to John, the Michigan lobby included shipbuilders Sheldon McKnight and Eber Brock Ward, Peter White of Marquette, and J. Vernon Brown, editor of the *Lake Superior Journal* in Sault Ste. Marie. From John Burt's room, which served as their headquarters, the men organized the plan for promoting the Soo Canal bill. Specimens of copper and iron were proudly displayed to congressmen like calling cards to emphasize the importance of the Upper Peninsula minerals.[98]

On August 11, 1852, Senator Alpheus Felch presented to the Senate a bill requesting either money or land to build the Soo Canal. Senator Cass emphasized the canal's military advantage rather than its value to commerce.[99] A bill was finally passed, and on August 26, 1852, it was signed by President Millard Fillmore. The state of Michigan was granted 750,000 acres of land to fund construction of the canal. The original proposal called for a transfer of only 500,000 acres of land, and it has been suggested that John and the other Michigan lobbyists may have been been responsible for the increased size of the grant.[100] Congress required that work on the canal project commence within three years, and a period of 10 years was allowed for its completion.[101]

Victory for the Soo Canal bill had not been easily achieved, nor was it entirely popular. In 1840 Senator Henry Clay had questioned the expenditure of any funds for the Upper Peninsula, an area he positioned "beyond the remotest settlement of the United States, if not the moon."[102] For many of the villagers at the Sault the portage business had brought prosperity and happiness which they believed would end with a canal at St. Mary's falls.

U.S. Senator Truman Smith of Connecticut insisted the canal could not be built in 10 years, if ever.[103] Even editor J. Vernon Brown, a member of the Michigan lobby, was pessimistic. He returned to Sault Ste. Marie and wrote an editorial in which he doubted there would be an early completion of the canal. John Burt responded to Brown by predicting that the project would be completed within two years. Historian William Chandler aptly wrote, "Mr. Burt was about the only one who had faith."[104]

Chapter XVII

1852 - 1855

*The use of [Burt's Solar Compass] has introduced a
new era into the public surveys and has rendered highly
interesting and scientific a business which before that time
was simply laborious.* [105]

-- John Wilson, GLO Commissioner
Feb 16, 1854

Confirmation on the Iowa-Minnesota Boundary

In 1852 GLO Commissioner Butterfield prepared his letter of support for William Austin's petition to Congress. He also took steps to confirm the value of Burt's solar compass by testing it on the survey of the Minnesota-Iowa boundary.

Although Congress had approved $30,000 to fund the project in 1849, it was not until February 2, 1852, that Commissioner Butterfield instructed Surveyor General George B. Sargent to issue contracts to establish, run, and mark the northern boundary of the state of Iowa.[106] On February 9th Captain Andrew Talcott of the U.S. Topographical Bureau was chosen as surveyor and astronomer to head the operation. Earlier, in 1833, Talcott led the boundary survey between Michigan Territory and Ohio during the Toledo strip dispute. He was assisted by a young West Point graduate named Robert E. Lee, who later commanded the Confederate armies during the Civil War.

The parallel of 43 degrees and 30 minutes (43° 30') north latitude was chosen for the Iowa-Minnesota boundary. It was to extend west from the main channel of the Mississippi River to the middle of the main channel of the Big Sioux River. Celestial observations were to be taken not more than 48 miles apart.

On March 2nd James M. Marsh was appointed examining deputy surveyor for the boundary survey. On May 12th he was informed that the Commissioner of the GLO was "desirous of testing fully Berts improved Solar Compass." Acting GLO Commissioner John Wilson told Surveyor General Sargent that "such a test, on so rare an occasion, is indeed a public duty."[107]

Captain Talcott outlined the two objectives of Marsh's survey. One was to obtain information regarding the nature of the country, and the other was to test Burt's solar compass. He informed Marsh:

> The Commission of the General Land Office deem it of great importance to the national Surveys, viz, to test the accuracy of a line surveyed by a solar compass in the hands of a Surveyor expert in the use of that instrument. The instrument you have is, I understand, one of the best and contains the latest improvements of the inventor and therefore is a proper Instrument to be used in this survey.[108]

Marsh was also instructed to run the preliminary line due west without reference to the lines that had been determined by the astronomical instruments used by Talcott:

> You will by the aid of the [solar compass] alone run as near as practicable a parallel of latitude.

The preliminary survey went smoothly for Marsh and his crew of 10 men, except that many of the supplies they had buried for later use were uncovered and taken by Indians.[109] On July 17th Surveyor General Sargent informed Commissioner Butterfield that Marsh's exploring line had reached the "Big Sioux" ahead of schedule. He wrote:

> You will ascertain that Burt's solar instrument has stood the test most admirably.[110]

Captain Talcott reported that the trial line run with Burt's solar compass was nearly identical to the line Talcott fixed by astronomical observations.[111] The only significant difference was the cost of running the two lines. Talcott's line cost over $32,000, or about $124 per mile. If the job had been done by Burt's Solar Compass, the cost would have been about $6,000 or $25 per mile.[112] It would appear that Burt's petition to Congress had been considerably strengthened, as the value of his solar compass was now fully confirmed.

Burt's Petition

John Burt's success as a Washington lobbyist extended beyond his role in promoting the Soo Canal bill. Prior to his arrival in Washington during December 1851 no action had been taken on his father's petition. Soon after John began talking to congressmen, however, his father's petition finally reached the Senate floor.

On February 6, 1852, Senator Felch announced to the Senate that William A. Burt was seeking "a just allowance in consideration of the benefit the government has derived from the use of a compass invented by him." Burt's petition was then referred back to Felch's Committee on Public Lands.

John then solicited additional testimonial support for his father to be presented to Senator Felch. On February 17th Lake Superior geologist J.W. Foster commented on his use of the solar compass:

> I unhesitatingly assert that I regard it as by far the best instrument now in use.. . . So thoroughly satisfied am I of the superior advantages of the solar compass over the magnetic needle, that I think that it ought to be required of the linear surveyors to employ it in all the public surveys, using the magnetic needle only for the purpose of determining the local variation.[113]

GLO Commissioner Butterfield informed the committee that "the solar compass had proven of inappreciable utility." He estimated that 31,104 miles, or nearly 10 million acres, of public lands had already been surveyed with the solar compass at a maximum cost of $186,624. He added that these lands "could not have been surveyed by other instruments for less than $622,080 and possibly a great deal more." In other words, the cost to survey the public lands with other instruments was "$20 per mile instead of $6, and possibly double the former amount."[114]

Butterfield pointed out that the solar compass was used in Oregon "as the only instrument calculated to insure anything like accurate results at moderate cost to the Government". He added:

> In fact, it is by the recommendations of the Surveyors General supplanting the use of the common compass throughout the public surveys, it being found the most serviceable instrument under ordinary circumstances even in Florida, where the variation is not great.[115]

Butterfield expressed his opinion that several hundred thousand dollars would be saved by using the solar compass on the surveys of the public domain. He recommended that a $15,000 or $20,000 allowance to Burt would be "a small consideration in comparison of the vast savings of public expenditures and the valuable & accurate results accruing from the use of that instrument in the surveying service."

John Wilson, Chief Clerk of the GLO, added his testimonial for Burt in a letter to Felch's Senate committee:

I have never seen or heard of its equal for the purpose for which it was intended. By specific instructions the United States surveys in the mineral regions of Lake Superior and in Oregon and California were required to be executed with it; and from the nature of the mineral lands on Lake Superior, it would have been impossible to survey them without it, unless at a cost to the Government exceeding the value of the lands. . . . Mr. Burt's invention stands among the foremost as entitled to consideration; it's utility has been fully tested and proved. His reward should be commensurate with the benefits derived from it. . . . A reward of one mill per acre for lands surveyed by his compass, which could not have been surveyed except at a cost of at least four times the amount paid for it [would be a] small measure of justice.[116]

But no action was taken on Burt's petition when the first session of the 32nd Congress adjourned in August 1852. There was still hope, however, that a bill could be presented during the second session scheduled to convene in December.

Preparations for the Soo Canal

Although Congress had authorized the land grant to fund construction of the Soo Canal, additional obstacles remained. A preliminary survey and estimate were needed in time for the next state legislature to approve the overall project. When Michigan's governor applied to the Secretary of War to send an engineer to draw up a plan and make the survey, he was rejected and told that funds had not been appropriated for that purpose.

During the summer of 1852, William Austin learned that Captain Augustus Canfield, a U.S. topographical engineer and son-in-law of Senator Lewis Cass, had volunteered to survey the route for the Soo Canal.[117] William Austin Burt would assist him in that important effort. The final details of the survey were discussed at Captain Canfield's home in Detroit.[118] In October 1852, with Sheldon McKnight providing their round-trip transportation, Burt and Canfield traveled to Sault Ste. Marie and completed the preliminary survey for the Soo Canal.

Following Congressional approval of the Soo Canal bill, John Burt worked to ensure its final approval by the Michigan legislature. Without his father's knowledge or permission, John promoted the nomination of William Austin as a candidate for the state legislature. He was elected as a representative from Macomb County.[119] John's friend Herman B. Ely was also elected to the Michigan legislature, as a representative from Chippewa County.[120]

On December 19th, several supporters of the canal project met with John to develop a consensus on several related issues, including the size of the locks. Under provisions of the Act approved by Congress, the locks were to be at least 250 feet long and 60 feet wide. The group concluded they should be at least 60 feet wide, and as long as possible.[121] Captain Canfield recommended, however, a 300 foot length for the locks, while J. Vernon Brown pressed for a minimum length of 350 feet.

As the state legislature was about to convene, a variety of several divergent plans for the canal threatened to indefinitely delay passage of the project. Congress required that construction of the canal must begin within three years and be completed within 10 years, or else proceeds from the land grant would be forfeited.

Michigan's Legislasture of 1853

Michigan's 17th legislature which convened on January 9, 1853, consisted of 31 senators and 71 representatives. William Austin Burt was chosen by the House members to head the Committee on the St. Mary's Falls Ship Canal Project.

Eber Brock Ward, owner of the largest fleet of ships on the Great Lakes, protested to Burt that building locks larger than 160 feet in length and 60 feet wide was both expensive and unnecessary.[122] Ward's views, however, were not shared by Charles T. Harvey, the 21 year old salesman employed by the Fairbanks Scale Company. Harvey had been recuperating from a bout of typhoid fever at Sault Ste. Marie when he heard that Congress had just passed the Soo Canal bill. He immediately wrote to his superiors, urging them to undertake the project of building the canal. Harvey, with youthful enthusiasm and the support of James F. Joy, a lawyer with the Michigan Central Railroad, sought to obtain the bid for the canal project.

In a bill presented to the House of Representatives on Jan 20, 1853, Harvey proposed the locks should be at least 350 feet by 70 feet. When William Austin Burt received Eber Brock Ward's letter urging smaller locks, he did not ignore his protest. Instead, Harvey was asked to appear before a special meeting of legislators from both chambers to defend his position on the larger locks. He was successful, and his recommendations prevailed.

On February 3rd William Austin Burt presented a final resolution recommending approval of the Soo Canal and it carried by a 51–10 margin.[123] The Soo Canal bill, signed by the governor on February 5th, provided for locks 350 feet long and at least 70 feet wide.

Michigan's 17th legislature finally adjourned after 41 days, having enacted 97 laws. On February 14th William Austin wrote to his son John in Washington, D.C., to share the good news of the canal bill's passage.[124] Regretfully, the railroad bill, strongly supported by John and opposed by the chartered railroad companies, was defeated in the Senate.

The canal bill authorized the governor to appoint five commissioners to prepare plans and to award the contract to build the canal. Details of this plan had been conceived at the December 29, 1852, meeting attended by John and other canal bill supporters.[125] In April 1853, the St. Mary's Falls Ship Canal Company, a subsidiary of the Fairbanks Company, was awarded a contract from the State of Michigan to build the canal. The terms required that the canal be completed by May 19, 1855; otherwise the contract would be invalidated. The deadline was eight years sooner than that required by the U.S. Government.

The First Taste of Victory

As the Soo Canal bill was being discussed in the Michigan legislature, John Burt was in Washington, D.C., to promote his father's petition. As yet no bill had emerged from either chamber of Congress to compensate William Austin for the government's use of his solar compass.

John was considered by some Detroit residents as a logical choice to become a commissioner for the Soo Canal project, but in a letter to his father he explained:

> My absence to Washington may be an objection, as it will be expected they will act soon after their appointment.[126]

To support his lobbying efforts John planned to show a Burt's Solar Compass to the Congressmen, much as he had done with the ore samples to promote the Soo Canal bill. This idea was thwarted when the instrument he had ordered from John Bailey in Detroit did not arrive. Bailey had been making Burt Solar Compasses for about three months with William Austin's blessing and supervision. Frustrated, John wrote:

> My business here gets along rather slow. I may go no further this session unless I get a Solar Compass soon.[127]

Confident that he would obtain "the appropriation" if he could display a solar compass, John teletyped William J. Young in Philadelphia to send one no later than February 15th.[128] Young complied, and at a "fair" John demonstrated the instrument to several Congressmen.[129]

On March 3, 1853, Senator Borland from the Committee on Public Lands presented a bill and favorable report:

> The Committee fully concurs in the views expressed by the Commissioner of the General Land Office, both as to the utility of Mr. Burt's Compass and as to the expediency and justice of granting some remuneration for the benefits which he has conferred upon the United States.[130]

The Senate approved the bill and it was sent to the House. Michigan's newly elected House delegate Hestor L. Stevens promised support for Burt's claim, but the 32nd Congress adjourned before it could act on the bill approved by the Senate.[131]

Railroad to the Upper Penninsula Iron Mines

Even without a formal charter Herman B. Ely continued his efforts to complete a steam railroad from the mines near present-day Negaunee and Ishpeming to Marquette Bay. On March 13, 1853, the Lake Superior Iron Company was formed on John Burt's land. The Lake Superior mine became known locally as the "Burt mine," with John Burt, his brothers, and Herman Ely the leading proprietors of the company.[132]

An Upper Peninsula traveler recounted his visit to the Lake Superior mine during the 1850's, while the railroad was being built. He called the location "Burt Iron Mountain," and reported:

> It has not yet been opened, yet those who understand such matters think it will pay richly to work it. The weight of [the ore] quite surprised us. We took hold of a piece about eight inches square and three in thickness, thinking to lift it with one hand, but our fingers slipped off as though it had been oiled, and no attempt was made afterward to lift any but the small pieces.[133]

Completing the railroad was an enormous undertaking. In addition to a shortage of labor and capitol, there were numerous obstacles, as described by Peter White of Marquette:

> The grades were heavy—rock cutting very expensive, bottomless swamps to cross, with fills numerous and difficult to make. . . . The severe winters of those days came with deep snows and extreme frosts, while the summer season brought overwhelming swarms of black flies, gnats and mosquitoes.[134]

In May 1853 John notified his brother William that " . . . the railroad is making excellent progress and they expect to complete five or six miles this season."[135] The following month John signed an agreement with Ely to extend the road two miles west of the Cleveland and Jackson mines to the Lake Superior mine, and to carry the iron ore at the same rate Eli was charging the other two mining companies.[136]

When the officers of the Cleveland and Jackson iron companies learned of Herman Ely's agreement with John Burt, they immediately broke their contracts with Ely and began constructing a plank road from their mines to Marquette Bay.[137] This action had been urged by an agent for the Cleveland Iron Company in December 1852. He was convinced that a plank road could be used throughout the year to benefit the community, while Ely's railroad would simply "fill the pockets of a few eastern men. . . ."[138] He added, "The only prospect of a railroad in my opinion is that it never will be built."

Ely responded with injunctions and lawsuits against the Cleveland and Jackson companies. After a fierce fight, causing considerable delay and expense in building the railroad, the conflict was finally resolved in 1855. Austin Burt and Charles T. Harvey served as arbitrators and, according to John, "our [railroad] came out victorious in the end."[139]

47.   Lake Superior Iron Mine No. 1
       courtesy Marquette County Historical Society

The Peninsula Iron Company

On August 28, 1854, the Burts and the Elys formed the Peninsular Iron Company with capital stock of $500,000. John Burt was joined by his brothers Austin and Wells and his father who had discovered the iron range a decade earlier. It was William Austin's first and only involvement in an Upper Peninsula iron mining company. Other stockholders included Herman Ely, and his brothers Samuel and George.

In March 1854 William Austin received a letter from Margaret Pilkington, with whom Burt had established a cordial friendship at London's Great Exhibition in 1851.[140] The message may have contributed to a decision by the Peninsular Iron Company and the Burt family to enter the blast furnace business.[141] Mrs. Pilkington wished to sell the timber from 600 acres of forest she owned in Chatham, Canada, across Lake St. Clair from Detroit. It would be a convenient source of fuel for a charcoal blast furnace, and eventually an agreement was made to purchase the wood.

Although economic conditions limited growth in the iron ore business between 1857 and 1863, the Civil War brought a new prosperity to the industry. In 1862 the Peninsular Iron Company sold 800 acres of its land holdings in the U.P. to the Lake Superior Iron Company. This enabled the company to purchase riverfront property along the Detroit River channel, in Hamtramck (now Detroit), where a blast furnace was constructed in 1863. The following year Austin Burt moved his family from Marquette to Detroit and became manager of the new furnace.[142] An important source of fuel was cord wood from the Pilkington land in Chatham, which was floated down river on scows and unloaded at the company's docks. John, William, and Wells Burt also took an active role in the Hamtramck furnace during its 25 year existence.

During the summer of 1872 John built a charcoal blast furnace on his property in the Upper Peninsula located at the mouth of the Carp River on the shore of Lake Superior. The following year the Carp River Iron Company was formed, with stockholders electing John president and his son Hiram vice-president. In January 1874 the company was consolidated into the Peninsular Iron Company. Soon after the furnace was put into operation an accident occurred.[143] A defective hearth in the furnace caused 10 tons of iron to spill out on the ground. Within three months, however, the problem was corrected and the furnace was operated sporadically until 1882.

One of 16 charcoal kilns reported to have supplied the Carp River furnace has survived. Built about 1878, it is called the Peninsular Iron Company's Mangum kiln. The historic site is located on Greenfield Road at Mangum Road in Gwinn, south of Marquette. One source describes it as:

20 feet in diameter at the base and 20 feet high, of rough coursed stone construction, built against a hill, with two openings arched with stone, each five feet high. One opening is at ground level, while the second is located on the side abutting the hill and is at the top of the kiln.

John was still president of the Peninsular Iron Company in 1884 when he announced the firm had completed its 30th year. John added, "This has been one of the most successful charcoal blast furnaces in this state."[144] At that time there were over 72 mines on the Marquette iron range, with total shipments exceeding 19.5 million tons of iron ore.[145]

The 33rd Congress

The first session of the 33rd Congress began in early December 1853. The following month northern Democrat Stephen Douglas introduced the Kansas-Nebraska bill that would render the Missouri Compromise inoperative. His action outraged northern senators, and overnight a wave of anti-slavery sentiment swept across the North.

Suddenly Congress was projected into the most turbulent period of its history. The bill, which became law in May 1854, kindled the passions of both North and South that would eventually lead to the Civil War. It also led to new political alliances. On July 6, 1854, at Jackson, Michigan, a group adopted the Republican name as an anti-slavery party and called for repeal of the bill.

48.   Peninsular Iron Company's Magnum kiln (c. 1878)
courtesy Charles K. Hyde

In January 1854 Michigan's Democratic Senator Charles Stuart introduced a new bill for William A. Burt's compensation. In his report, Stuart said:

> The Committee on Public Lands] are well satisfied that he is entitled to a reasonable compensation from the United States for the great benefit resulting from the use of his Solar Compass.

The following month Stuart asked the Senate to appropriate $15,000 for Burt without debate. One Virginia senator voiced concern that approval of this bill to compensate Burt would set a dangerous precedent. Stuart, however, reminded the senators that a much larger appropriation had recently been granted without debate to three Army officers for their inventions relating to percussion caps.[146] It was a reasonable argument, but military personnel were apparently regarded by a different set of criteria.

In fact, during the autumn of 1854, Minnesota surveyor Thomas Simpson met two Army officers who had just completed the survey of a military road from Sioux City to Fort Snelling. Impressed with a solar compass Simpson had shown them, one officer remarked it was a shame the instrument had not been introduced into the Army for use by the engineers. Simpson recalled, "The only reason he could give was that it had not been invented by an Army officer."[147]

Senator Lewis Cass supported his Michigan colleague and praised the solar compass:

> This invention we have used; our officers have used it in the surveys; they could not have done their work properly without it. It is a beneficial and most useful invention; we have used it, and saved thousands and thousands of dollars, as well as disputes with regard to lines; and now the recommendation to is to pay him the small sum of $15,000. I cannot see upon what principle you can refuse it.[148]

Stuart, noting that Burt had not patented his "improved" invention, assured the senators that if it had been his decision he would have asked for a new patent rather than take ten times the amount that was being requested. It was an open admission that perhaps Burt had been ill-advised to allow his patent to expire.

On March 10, 1854, the bill "to enable the United States to make use of the solar compass" passed in the Senate. It was nearly defeated by a block of nine Southern senators, a pattern that would be repeated. Once again, however, the bill failed to reach the House before Congress adjourned.

PART FOUR

Chapter XVIII

1855

He that has patience may compass anything. [1]

- Francois Rabelais (1490 - 1553)

Completion of the Soo Canal

In early 1855, John Burt was much too busy with his own affairs to concern himself with his father's case before Congress. On February 15th, three days after Michigan's railroad bill finally passed the state legislature, the Iron Mountain Railroad was incorporated. John became its first president. [2]

It was also in February, with work on the Soo Canal and locks nearing completion, that Michigan's Governor Bingham selected John Burt to become the canal's first superintendent. Marquette's Peter White said that the "important services" John had rendered during construction of the canal contributed to this appointment. White also acknowledged that federal approval of the land grant for the Soo Canal was largely due to John's persevering efforts. [3] As superintendent John Burt would oversee the local operations of the canal at an annual salary of $1,500 for a two-year term.

Governor Bingham's decision to appoint John canal superintendent delighted many people, although not all. J. Vernon Brown had campaigned for the job. A petition signed by 30 men had been submitted in support of Brown, but John had the backing of John W. Brooks, vice-president of the Canal Company who considered Burt to be fair and honest. [4]

Brown, as editor of the Lake Superior Journal at Sault Ste. Marie, was highly critical of Charles T. Harvey and other officers of the Canal Company. In an editorial on August 19, 1854, Brown expressed concern about the "slipshod manner in which the embankment was being done,"and he predicted it would be shown the locks were poorly constructed. [5]

Harvey was, in fact, racing against the May 19th deadline. If the canal was not built by that date, the Canal Company's contract would be invalidated. In February Harvey notified his superiors that one wall of the canal contained a weakness from a drainage opening, and time was needed to allow the fill to settle. [6]

John officially became superintendent on April 1, 1855, and on April 19th water was let into the canal. On May 21st, control of the Soo Canal passed to the state of Michigan, and John formally accepted it. [7] The project had taken over 22 months to complete and cost slightly less than $1 million. The canal was 5,700 feet long, 100 feet wide at the surface, and 13 feet deep. Each of the two locks were 70 feet wide, and 11.5 feet deep, with a lift of about nine feet. [8]

On June 18th the steamer *Illinois* was the first ship to enter the Soo locks, crossing from Lake Huron up into Lake Superior. Two months later the brig *Columbia* carried the first cargo of iron ore from Marquette through the Soo locks to Cleveland. As John watched the *Columbia* pass by, he undoubtedly recalled with satisfaction the promise he made to Dr. Peter Shoenberger two years earlier. [9]

It was a proud moment for John and a momentous achievement for Michigan. Fifty years later Marquette's Peter White remarked:

> The opening of the Sault Canal has been one of the largest benefit to the whole United States of any single happening in its commercial or industrial history.

49.   Early construction of the lock pit for the Soo Canal
       courtesy Marquette County Historical Society

50.   Completed Soo locks of 1855 (*author's photo from A.O. Backert, THE ABC OF IRON AND STEEL (Penton Publishing Co., 1921, 4th ed.)*

The Lake Superior Mining Journal

The rapid population growth in the Upper Peninsula's iron region created the need for improved communication. Until 1854, there was no mail service during the winter to Marquette County. Peter White recalled:

> I have known intervals of three or four months at a time when no mail, or letter, or news of any kind was received by anyone in the county.[10]

In 1855, to satisfy this need, John purchased the *Lake Superior Journal* at Sault Ste. Marie from J. Vernon Brown. The newspaper had been on the market for several months at a listed price of $3,000.[11] After moving the newspaper to Marquette, John changed the name to *Lake Superior Mining Journal*. It was, according to John, "the first paper to represent the iron interest of Lake Superior, as well as the first successful paper of any kind published in that region."[12]

John maintained ownership of the newspaper until 1859 when he sold it to Warren Isham, former editor of the *Michigan Farmer*. In 1885, an article in the *Mining Journal* praised John Burt as the newspaper's "foster-father".[13]

The Solar Compass is Challenged

Although William Austin Burt had not yet been compensated by Congress for the government's use of his solar compass, his invention was becoming widely accepted among surveyors. In 1854 several Illinois engineers challenged Leonard Neitz, a St. Louis mathematical instrument maker, to demonstrate the superiority of the solar compass over the common magnetic compass.[14] After Neitz ran an accurate meridian line with a Burt Solar Compass he had built, the engineers bet him $20 he could not precisely retrace the line.

Neitz won the bet and proceeded to demonstrate how Burt's invention could be used to sight on a distant tree with much greater accuracy than was possible with the magnetic compass. Convinced, the engineers gave Neitz an order for two solar compasses.

Burt & Bailey

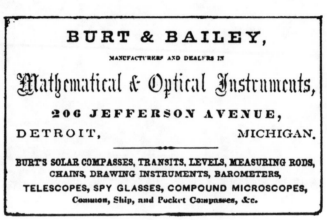

51.

courtesy Marquette County Historical Society

By allowing his patent to expire in 1850, William A. Burt also relinquished control in the development of his solar compass. Instrument makers were free to devise their own improvements, and often did. In 1852, however, William Austin arranged for John Aylesworth Bailey, a Detroit mathematical instrument maker since 1850, to construct Burt's Patent Solar Compasses under the inventor's personal supervision. In September Burt praised Bailey in a letter to GLO Commissioner John Wilson:

> John Bailey is a common genius, a very industrious man. He is now making a first rate solar compass and will soon be able to make four per month.[15]

Bailey had completed five solar compasses and was selling them at the competitive price of $150 each.[16] By July 1853 William Austin's son William formed a partnership with John Bailey under the name of Burt & Bailey to make and repair mathematical instruments including Burt's Patent Solar Compass. Inclusion of the words "Burt's Patent" was stamped on each solar compass to add credibility to those instruments personally authorized by Burt.

In December 1854 a new partnership agreement was formed when John Burt joined Burt & Bailey. He and William each held a one-quarter interest. Within a year, however, John Bailey withdrew from the firm to "engage in other business."[17] Until 1857 John & William Burt continued to make mathematical instruments at the same location under the name of J & W Burt.

William Burt continued the business with John Watson under the name of Burt & Watson until October 1858 when the business was sold to Charles Crosman and William C. Grant, two employees since 1853.[18] Two Burt and Watson solar compasses are displayed at the Henry Ford Museum in Dearborn, Michigan. Each are engraved with "1857-1858," and "Burt Solar February 25, 1836 Patented."[19] The firm of Grant & Crosman continued to make Burt's solar compasses and other mathematical instruments until at least 1861.[20]

52.   Burt's Solar Compass made by John Bailey in 1852, with sun's rays concentrated
        between crosshatched marks on silver plate below lower lens.
        Courtesy Stan Weaver (author picture)

The Manual of Instructions

The expanded government surveys created the need for a uniform set of surveying instructions to regulate the field operations of the deputy surveyors. On February 22nd the 1855 Manual of Instructions to Regulate the Field Operations of Deputy Surveyors was issued by the General Land Office. Over the years this manual has been revised several times, but the basic system of rectangular surveying has remained essentially unchanged.

The 1855 Manual was an expansion of the 1851 Oregon General Instructions that required the use of Burt's improved solar compass or an instrument of equal utility in areas where the variation of the needle was not uniform. It was "deemed best" to use the solar compass "under all circumstances." By an Act of Congress approved May 30, 1862, the Manual of Instructions became an integral part of every contract for surveying the public lands of the United States.[21] The solar compass, therefore, came into "general use" in the U.S. public land surveys in 1855, with the first Manual of Instructions to apply to all U.S. deputy surveyors.[22]

The "general use" of the solar compass did not entirely preclude the use of the magnetic compass. In 1944 Illinois surveyor W. D. Jones, commenting on the original instructions governing public land surveys, wrote:

The 1846 instructions, and all after them require "Base, meridian, correction and township lines . . . to be run with an instrument that operates independently of the magnetic needle." But up to and including those of 1855, the instructions say, "Where the needle can be relied on . . . the ordinary compass may be used in subdividing and meandering." We sometimes think that the surveyors had too much faith in the needle.[23]

The Manual of 1890, however, prohibited use of the magnetic compass except in areas free of local attraction, and then only for subdividing and meandering. The magnetic instrument was totally eliminated for use on the public land surveys beginning with The Manual of 1894.

To make a profit under the contract system of surveying, surveyors occasionally chose methods that would only minimally meet acceptable standards.[24] The added cost and sophistication of the solar compass compelled many contract surveyors to stick with the magnetic compass unless their work absolutely required the solar instrument. In 1879, U.S. deputy surveyor John Fitzhugh, who worked in Curry County, Oregon, cautioned:

> The solar compass is a fine instrument in the hands of an astronomer and mathematician. When in perfect order its results are quick and reliable, but under the control of an "ignoramus" it is the wildest "machine" that ever was used to trace a line.[25]

Nathan Butler, a U.S. deputy surveyor in Minnesota for 30 years, however, called the solar compass "the most convenient and efficient instrument ever used in surveying the government lands."[26] Butler added:

> The solar compass has the reputation, among a certain class of men, of being imperfect and unreliable. It has done poor work, no doubt, but that was not the fault of the compass,— rather of the man who was using it.

In his review of early instruments in the history of surveying, Robert P. Multhauf of the Smithsonian Institution concluded:

> [The solar compass] . . . can be taken to mark the end of the period when the great majority of surveyors proceeded about their necessary business relatively independent of the contemporary currents of science and technology.[27]

William A. Burt, Author

Soon after the General Land Office issued its 1855 Manual of Instructions, William A. Burt completed his own pocket-sized guidebook entitled "*A Key to the Solar Compass and Surveyor's Companion.*" Unlike the usual textbooks, Burt's leather-bound work is based on his 17 years of field experience and contains over 200 pages of practical information to assist the novice surveyor on a lengthy survey in the wilderness. As William Austin explained:

> The inexperienced surveyor in this branch of the public service has need of all the necessary information to enable him to accomplish his arduous duties in a proper manner.

He considered the lack of such practical knowledge a principal cause of inaccurate or fraudulent survey work.

Burt's manual contains tips for outfitting a survey crew of six men for four months in the field. Included are optimum quantities of various provisions, such as :

- 8 barrels of flour

- 8 bushels of beans

- 2 bushels of dried apples

- 120 lbs. good dry sugar

- 70 lbs. of ground coffee

- 25 lbs. of rice

- 4 lbs. of Castile soap

For "wearing apparel" Burt advocates a wool hat is best, and a light coat should contain water-proof pockets to keep field notes dry. Flannel, he says, is preferable to cotton for underwear in "all seasons and kinds of weather." A large silk handkerchief, any color except red, should be taken along to tie over the ears and neck as "good protection from flies and musquitoes". A supply of 3 ounce tacks should be carried for nailing boots. Burt also describes how to build a raft within two hours that can carry an entire survey party across a lake or river.

He advises how to use, adjust, and even make the solar compass, but he cautions that the parts should meet certain specifications. Accurate results cannot be expected if care is not taken in its proper construction and adjustment. During 1855, when a potential manufacturer of solar compasses offered to pay William Austin a royalty for each instrument he made, Burt replied:

> . . . as to the fee for the right to make solar compasses, it is only to make them according to the plan laid out in the book. I have no other charge to make.[28]

Burt mailed a copy and billed him $2.00, satisfaction guaranteed.

The first edition of *A Key to the Solar Compass and Surveyor's Companion* was published in 1855 by William C. Young, in Philadelphia.[29] Seven subsequent editions were published by D. Van Nostrand, of New York, through 1909. The utility of Burt's practical guidebook is evidenced by the fact it remained in print until 1929, a period of 74 years. A facsimile by Carben Surveying Reprints is currently available through Landmark Enterprises in California.

53.   Burt's Equatorial Sextant - 1856 *courtesy Smithsonian Institution*

Chapter XIX

1856

If one advances confidently in the direction of his dreams, and endeavors to live the life which he has imagined, he will meet with a success unexpected in common hours. [30]

- Henry David Thoreau (1817 - 1862)

The Equatorial Sextant

Since 1850 William Austin had tinkered with the idea of modifying his solar compass for use by ships at sea. While crossing the Atlantic from London's Great Exhibition to America in 1851, he observed the need for more accurate instruments to determine a ship's precise location.

By early 1855 the first prototype of his new instrument to meet this need was constructed by Burt & Bailey in Detroit.[31] It consisted of two principal rings fitted together. The inner "latitude circle" could be rotated 90 degrees inside the outer "meridian circle" to which an "azimuth circle" was attached. The ship's position was determined by setting any three of the five functions of the astronomical triangle: declination, latitude, altitude, hour angle, and azimuth.

On April 5, 1855, Burt submitted his plan to William J. Young who promptly replied:

> You ask me my opinion as regards the new astronomical or nautical inst. you have invented. It will not take many words for an emphatic answer. If you do succeed in planning an inst. which can readily be used, and not easily put out of order, I would like to take an interest in it. . . . The main thing to be attained is the meridian or variation of the needle. If that can be accomplished, every steamer *must* have at least one.[32]

Young intimated that perhaps Burt was attempting to accomplish too much with his nautical instrument. William Austin nevertheless continued with his planned design. In August 1855 Burt informed Young that he had obtained a variation reading on board the *Illinois,* possibly during its historic first journey through the Soo Canal on June 18th. Young acknowledged Burt's concern about the cost of the instrument:

> . . . it will be always, of necessity, an expensive inst.. The utility in many cases will justify the expense; for I think you would have found the variation much greater if on board an iron steamer or a vessel loaded with rail road iron. A few voyages in such cases would amply repay the owners for the cost, and *they ought* to pay for it, if notmthe captain. I should suppose most of the government vessels would have one.[33]

Young reiterated that the main purpose of Burt's new instrument should be to obtain the variation. The original model was accurate to within five degrees, but Young was confident that a more efficient and less expensive instrument could be made to an accuracy within one degree.

Burt considered several possible names for his new nautical compass, including "Marine Solar Compass", "Computing Sextant", and "Pantalobe."[34] He chose, however, "Equatorial Sextant", as the original compass was combined with a reflecting sextant. Later, this arrangement was found too cumbersome.

In March 1856 Burt sent one of his Equatorial Sextants to the Navy Department, requesting an examination and report. Lt. Matthew Fontaine Maury, noted superintendent of the U.S. Navy Observatory and Hydrographical Office, was sufficiently impressed with Burt's invention to recommend it to his superiors.[35] Referring to the Equatorial Sextant, Maury wrote:

> It is very ingenious and beautiful in principle, and has great compass. It is capable of solving a problem of this sort.[36]

It was also in March 1856 that one of Maury's professional rivals, A. D. Bache, then Superintendent of the U.S. Coast Survey, also evaluated an Equatorial Sextant sent to him by John Burt.[37] Bache concluded:

> . . . the principles are correct, and when the instrument is accurately made and adjusted it will probably give a degree of precision corresponding to that attainable in the observation of the altitude.[38]

It was one of the few times that Bache and Maury concurred with each other on anything. Bache suggested, however, that the instrument be made lighter for use on ships and recommended separating the sextant and computing parts.

Encouraged by these reports, William Austin traveled to Philadelphia in August 1856 to supervise William Young's construction of three Equatorial Sextants.[39] After Burt delivered the instruments to the Navy in Washington, he reported:

> My Equatorial Sextants were accepted by Lieut. Maury as answering the purpose designed in ordering them.[40]

It was an important first step.

On November 15, 1856, U.S. patent 16002 was issued to William Austin Burt for his Equatorial Sextant, and it was reported in the *Scientific American*. The following year foreign patents were received from France, England, and Belgium.[41] William Austin, aged 64, was suddenly dreaming that someday every ship might be equipped with an Equatorial Sextant. Clearly he foresaw greater worldwide potential for his nautical instrument then his solar compass.

The Compensation Debate Continues

As the 34th Congress convened in December 1855, political unrest had reached a volatile stage. The House was in chaos and so torn by factionalism that it required two months to elect a Speaker on the 133rd ballot. Even then it lacked a majority, and for the second time in U.S. history the Speaker of the House was chosen by a plurality.

Such was the climate in Washington, D.C. when Michigan Senator Charles Stuart introduced a new bill requesting $15,000 to compensate William Austin for the government's use of his solar compass. With little discussion the bill was approved and sent to the House.

On March 31st Representative David Walbridge introduced in the House a separate bill "to make use of the solar compass in the public surveys." The title of the bill presented by Walbridge was misleading, and Democrat George Washington Jones of Tennessee inquired:

. . . if I understand the matter aright, the bill is a proposition that the Government shall buy the patent right. I ask the gentleman if this Government has ever furnished compasses to any persons who have contracted to survey the public lands?[42]

Walbridge explained that the inventor was simply asking a fair compensation for the millions of dollars the government had saved and would save in the future by the use of the solar compass. He added:

The Government insists that [deputy surveyors] use this very instrument. . . . Therefore, they compel the deputy surveyors to use it; and they buy them of the manufacturers, at the manufacturers cost prices, and nothing is paid to the inventor for his patent at all.

Following a brief discussion the bill was referred to the whole House.

The following month Congressional tempers surpassed the breaking point. Massachusetts Senator Charles Sumner, a strong advocate for abolition, delivered a stinging attack against the administration's pro-slavery position for Kansas. His tirade included vicious statements against two Senators, including Andrew Pickens Butler of South Carolina who was out of Washington and unable to respond to Sumner's remarks. Butler's nephew, South Carolina Representative Preston S. Brooks, however, did not take the remarks lightly.

On May 22, 1856, two days after Sumner's speech, Brooks and South Carolina Representative Laurence M. Keitt entered the Senate chamber as it adjourned, both carrying gutta-percha canes. Sumner was still writing at his desk as Brooks walked up behind him and began beating his head with the cane until Sumner fell unconscious. Keitt stood nearby, preventing other Senators from interfering with the attack. The anti-slavery forces in the House were incensed by the incident, but a motion to expel Brooks and Keitt was defeated. Both men resigned, but they were immediately re-elected by their constituencies.

It was two years before Sumner could return to the Senate, but he became a martyr among the Republicans. Brooks and Keitt became heros among their Southern colleagues. By July 26th they had re-occupied their seats as the House took up the bill for William A. Burt's compensation.

Senator Walbridge urged approval of the bill. He explained that by an agreement between Burt and the GLO Commissioner, the inventor did not accept any royalties from U.S. deputy surveyors, only county surveyors or those not under contract with the government.[43] In return, the Commissioner would recommend that Congress compensate Burt for the use of his compass. Although William Austin usually waived the royalty fee it is unlikely any such agreement took place prior to December 1849.

The Representative from Missouri argued that the U.S. Government was not bound to furnish deputy surveyors with any instruments whatever. The Commissioner, he believed, had no authority to instruct the contract surveyors to use the solar compass. His contention was only partly valid. The first official Manual of Instructions directed from the GLO Commissioner to all U.S. deputy surveyors was not published until 1855. Earlier each Surveyor General issued instructions independent of the GLO or other surveyors general.[44] Nevertheless, beginning with Michigan in 1840, individual surveyors general required the use of the solar compass on selected survey lines. In those instances the contract surveyors could not use other instruments.

When Walbridge informed Representative Jones of Tennessee that Burt had not been remunerated for his solar compass, Jones replied:

That is his own loss . . . Sir, the principle is wrong that the government should, in its governmental capacity, be the rewarder of all those who discover useful and important inventions. . . . What, sir, has the government paid to Mr. Colt, the inventor of the revolver . . . What remuneration, sir, has the Government made, or should make to him? When the Government needed the arm which he had invented, and which he had the exclusive right to dispose of and control the Government bought it at his price and used it.

Ironically, Samuel Colt received his first U.S. patent for a revolving fire-arm on February 26, 1836, the same day that William Austin Burt received his patent for the solar compass. In 1842, when his business failed, Colt sold his patent rights to others. It was not until the Mexican War broke out and the U.S. Government ordered 1,000 Colt revolvers that Colt bought back his patent rights. Two years before the patent was due to expire he applied for and received an extension of seven years. In 1854, however, a House select committee investigated charges alleging Colt had engaged in bribery to secure his first patent extension. Although the charges were not entirely supported, there was no doubt that Colt knew how to muster Congressional support. Now, in 1856, Colt was seeking an additional extension of his patent.[45]

Michigan Representative Peck responded to Jones:

> I answer, the Government has rewarded Mr. Colt, and the people who have used Mr. Colt's invention have rewarded him most bountifully. . . . But in this case it is far different. . . . In this case the invention is required only in a limited sphere . . . Let the enjoyment of the rights guaranteed to inventors under the patent laws be their remuneration, if that be sufficient; but if that be not sufficient, and if the invention be a valuable accession to the cause of art and science, and if, in addition to that, it is absolutely essential to the discharge of important functions of the Government, than it is the duty of the Government to reward the inventor.

Peck then appealed to the Representatives' sense of justice:

> [William A. Burt] is not of that class of men who invent revolvers and make money by the operation. He adds to the treasures of the country, and not to his own.

On a motion from Georgia Representative Alexander Stephens debate on the bill was ended before a final vote. On December 12, 1856, the bill recommending $20,000 for Burt was again brought before the House, with Walbridge recommending passage. The Senate bill to compensate Burt $15,000 was also on the House calendar.

Walbridge reminded the Representatives:

> The patentee acted under the advice of the Commissioner, and, as advised, did permit the Government surveyors to use his instrument, expecting that it would, in the end, remunerate him for that use.

One Southern Representative reiterated earlier concern that passage of the bill would set a dangerous precedent, and before Walbridge could respond, a motion by Jones of Tennessee to table the bill was approved. Supporting that effort was a solid block of Southern Representatives including both of Senator Sumner's attackers, Brooks and Keitt. Curiously, 28 representatives from the large industrial states of New York, Philadelphia, and Ohio also voted to table the bill.

The Senate-approved bill never reached the floor of the House. A golden opportunity for Burt's overdue compensation was lost, and it was evident that new strategy was required.

Chapter XX

1857 AND 1858

The sun and moon and stars keep pace with
him;
Invisible hands restore the ruined year,
And time, itself, grows beautifully dim. [46]

- David Morton (1886 - 1957)

John at the Soo

Surprisingly, traffic at the Soo was not as heavy as anticipated during its first season. In 1855 John Burt collected less than $4,500 in tolls, and only about 109,000 tons of cargo passed through the canal gates.

Part of the reduced traffic was attributed to the difficulty some ships had in navigating the rapids and bends in the St. Mary's River, causing a delay of nearly two weeks.[47] To solve the problem, John offered an incentive to steam tugs assisting vessels traversing the difficult portions of the river. Tugs passing through the canal with vessels in tow were not charged a toll. The idea was successful, resulting in increased traffic and revenue. In 1856 iron ore shipments passing through the canal from the Upper Peninsula mines increased in tonnage from 1,447 to 11,597.[48]

The safety measures established by John during his first winter season as superintendent were generally adopted by his successors.[49] The locks were partially filled with water and the gates were closed. Measures were also taken to prevent ice from accumulating at the lock gates that could cause serious damage from expansion.

When frost eventually formed along the earth embankment, John feared a rupture if the level of Lake Superior rose suddenly during the spring run off. In his November 30, 1856 report to the Board of Control, John suggested strengthening the embankment by constructing a stone wall outside of it, widening it, and extending the north pier at the upper end. The problem had not been resolved when John's two year term ended on April 1, 1857. This may have been a factor in the selection of John's successor, his brother-in-law Elisha Calkins.

Born in Cayuga, New York, October 1, 1816, Elisha Calkins came to Michigan with his family in 1825, settling in Washington Township near the Burts. Ten years later John Burt married Elisha's older sister Julia. Like John, Elisha embraced the Republican party soon after its formation. Prior to his appointment to succeed John as superintendent of the Soo Canal, he was a supervisor of Washington Township for five years.[50]

Although it is strictly conjecture, the selection of Elisha Calkins as superintendent may have permitted John to retain a strong voice in decisions involving the canal. Clearly, however, Calkins was not just a figurehead. In his annual report to Governor Bingham dated December 28, 1857, Calkins suggested building a stone wall inside the canal embankment with "timber or fender work," and noted:

> I am compelled to differ somewhat from my predecessor in his recommendations upon the same subject.[51]

54. Soo Canal Lock (1870's) *courtesy Chippewa County Historical Society*

It is extraordinary, however, that Calkins' annual report for 1857 was never included in official records. It is excluded from the bound edition of the Reports of the Superintendent of the St. Mary's Falls Ship Canal, 1856—1873 as well as the Joint Documents of the State Legislature. An annual report for 1857 is included in these volumes, but is is written and signed by John Burt. Burt's report, written on stationary bearing the letterhead "Superintendent of the Saint Mary's Falls Ship Canal," was dated December 20, 1857, nearly nine months after John's term of office had expired.

On December 12, 1857, Governor Bingham had called a special session of the Michigan Legislature to consider the urgent need to improve the canal. He had not received Calkins 1857 annual report by mid-December when Bingham asked John Burt to furnish him with all the facts known to him "relative to the condition and safety of the Saint Mary's Falls Ship Canal." John's comprehensive reply on December 20th included a recommendation for an additional set of locks for safety and management reasons, to be financed with Congressional funds.[52]

The legislature was unsuccessful in their request to Congress for $50,000 to repair the Soo canal, but eventually state bonds were issued to complete the work. Although Elisha Calkins first annual report dated December 28, 1857, was filed with Governor Bingham, it remains a mystery why John's unofficial report was the one that has been included in the official records.

Completing the Upper Peninsula Railroad

During John's term as superintendent of the Soo Canal, he maintained his interest in the completion of the Upper Peninsula's first railroad. In 1856, while John was president of the Iron Mountain Railroad, the company obtained the necessary funds from New York and Boston capitalists to complete the line.[53] Unfortunately, Herman B. Ely, the railroad's prime mover, died before it began operating in 1857. Samuel P. Ely, however, assumed leadership of his brother's business activities.

In December 1855, John traveled to Washington in a successful effort to obtain a land grant for a second rail line. In John's words:

> I log rolled in Congress on my own motion and responsibility to obtain a land grant to aid in the construction of a R. R. from Little Bay de Noquette to Marquette and thence to Ontonagon, which act passed June 3, 1856. . . .[54]

The Bay de Noquette and Marquette Railroad Company was formed on November 21, 1856, and in February 1857, John Burt was chosen its first president. By 1858 the line was extended to the Lake Superior Iron Company (Burt Mine) in Ishpeming, and in April 1859 the railroad purchased the Iron Mountain Railroad Company. During 1863, however, John sold his stock and severed all ties with the Bay de Noquette and Marquette Railroad Company. Nine years later it became part of the Marquette, Houghton, & Ontonagon Railroad, which was consolidated with other railroad lines. Eventually it evolved into the Soo Line Railroad System.

Time Runs Out

In 1857, after 32 years in Macomb County, William Austin and Phebe bid farewell to friends in Mt. Vernon and moved to Detroit. Their sons William and Austin had already established residency in Detroit, so more time could be spent with their grandchildren. The move also made it possible for William Austin to supervise more closely the construction of his Equatorial Sextants at Burt & Watson. Wells Burt remained in Macomb County, while John continued his activities in the Upper Peninsula.

At age 65 most men settle down to a leisurely lifestyle, but William Austin still had some irons in the fire. The matter of his federal compensation was still unresolved. It had been seven years since he first petitioned Congress, and the prospects for success grew dimmer with each passing year. On January 1, 1857, Burt petitioned Congress to renew and extend the solar compass patent for 14 years, in lieu of the

55.   Phebe Burt   William Burt
       (author's collection) (Phebe print from portrait in Marquette County Historical Society;
       WAB cut from Pratt, LIFE AND TIMES OF HENRY BURT)

small allowance previously requested.[55] He affirmed that he had never received as a compensation, either directly or indirectly, more than $300 for his right in the solar compass "or the patent thereof." The petition was not introduced until the following year.

Most of William Austin's attention during this period was focused on developing and introducing his Equatorial Sextant for widespread distribution. He prepared a manual entitled *Description and Use of the Equatorial Sextant* that would accompany the sale of each unit sold.[56] By late October 1857 Burt informed Charles Stansbury, a Washington, D.C., patent agent, that patents had been received from England, France, and Belgium for his new invention.[57] Three months later Burt sent Stansbury two Equatorial Sextants (the original instrument made by Burt & Bailey and one by Burt & Watson). The latter instrument was made without sextant glasses and was of the size and form that Burt proposed to use for future production. When the observing parts were required the Burt & Bailey instrument was to be used as a guide.

William Austin then instructed the agents to be governed by their own judgment as to the best interest of the patentee in regard to selling the right to make Equatorial Sextants. He was to be kept advised on sales progress, and the agents were to receive a 20% commission on all sales they collected.

On February 4, 1858, Dewitt Leach, a U.S. Congressman from Michigan, introduced in the House of Representatives Burt's memorial for a 14 year renewal or extension on his solar compass patent. Although it was referred to the Committee on Patents it was never taken up for discussion.

In mid-March, while instructing a group of 12 sea captains on the elements of nautical astronomy and the use of his Equatorial Sextant, William Austin was stricken with a severe heart attack. During the next five months he endured extreme suffering.[58] His heart condition often prevented him from lying down, and much of the time he was confined to a chair.

136

56.   Burt Monument, Elmwood Cemetery, Detroit. *(author collection)*

The optimism that had carried William Austin Burt through so many previous ordeals was suddenly gone. A letter he received in April from William J. Young's son Alfred went unanswered.[59] In May, William Young sent Burt a detailed sketch for a compact method of attaching a telescope to a solar compass.[60] Although Burt's opinion was solicited it was never received. At 6 P.M., Tuesday, August 18, 1858, William Austin Burt died at his Jefferson Avenue home in Detroit.

Horace Burt, Austin's son, rode on horseback from Detroit to Mt. Vernon to carry the news of William Austin's death.[61] Funeral services were held in the Baptist Church he had helped to form, across the street from his former home in the Washington area of Macomb County. He was buried in the cemetery next to the church. Six years later, following the death of his wife Phebe, his remains were moved to Detroit's Elmwood Cemetery, final resting place for many of Michigan's prominent citizens.[62]

57. Marquette in 1881, showing breakwater
      courtesy *Marquette County Historical Society*

Chapter XXI

THE BURTS IN THE UPPER PENINSULA

*The value of life lies, not in the length
of the days, but in the use we make of
them...*[63]

- Michel De Montaigne (1533 - 1592)

Marquette: Sparks of Growth and Disaster

In the decade following their father's death, John, Austin, and William Burt took an active part in the growth and development of Marquette. Wells maintained his financial interest in the Lake Superior and the Peninsular iron companies, but he preferred to remain in the Lower Peninsula for health reasons. In Mt. Vernon and later Ypsilanti he could avoid the pressures of an active business life.[64] In 1873, however, Wells moved to Detroit to help form the Union Iron Company. He served as its president for 10 years.

Marquette was incorporated as a city in 1859. The county, town, and iron range are named after the French Jesuit missionary Jacques Marquette, who explored the Lake Superior area in the late 17th century. By 1860 Austin Burt and his family had moved from Detroit to Marquette. John was already well established as a prominent Marquette citizen and owned a large portion of the business district.[65] William did not become a permanent resident of the city until 1866.

During 1858 John and his brothers formed the Burt Brothers sandstone quarry, which was located just south of Marquette. A small stone structure, known today as the "John Burt House", was built to serve as a warehouse and clerk's office. Restored in 1955, the attractive pioneer home is maintained by the Marquette County Historical Society as an Upper Peninsula tourist attraction during the summer months.[66] With thick walls and a small attic window it was locally called "The Fort," although it was never used for military purposes. Included are a storage cellar, two downstairs bedrooms, and an attic.

The Marquette sandstone was considered a highly valuable building material, and the Burt quarry soon gained a national reputation.[67] In the Spring of 1861 the Burts sold the quarry to George Craig and his son Thomas, but in 1872 they purchased it back under the name of the Burt Freestone Company. Incorporated in October 1872, the principal members of the new company were John Burt, his son Hiram, William Burt, and his sons A. Judson and William Austin.

The Burt brothers engaged in several business ventures in Marquette, but in the business district none was more visible than the "Burt block." It was the largest building in town, a three-story structure of brick, with stone trimmings, which was completed in 1864.[68]

John's sons Hiram and Alvin, in their mid-twenties, operated Burt Brothers' General Store on the main floor of the "Burt block." The store carried a broad assortment of general merchandise as well as mining supplies and logging machinery. Peter White's First National Bank was also on the street floor, along with Stafford's Drug Store. Evan Brothers' barber shop and bathrooms were located in the basement of the building to provide customers with a shave and, for an extra quarter, a hot bath in a copper-lined tub. Few homes were equipped with bathtubs in the early 1860's. The top floor of the structure contained law offices and the business offices of the Michigan Iron Company and the Marquette & Ontonagon Railroad.

58.a   Wells Burt   58.b Austin Burt
courtesy of John H. Clarke (author file)

Ever since the late 1850's John had been an active member of the Baptist church in Marquette. The congregation was without a permanent home until 1862 when John Burt donated land; and a building, largely financed by him, was erected.[69] Since 1937 the site has been occupied by the Marquette County Historical Society museum and library, where the Burt Papers are located.

By 1864 an urgent need had developed for a breakwater to protect the Marquette harbor. One resident counted as many as 62 ships anchored off shore, waiting their turn at the ore docks owned by the mining companies, and hoping for calm weather.[70] Work on the 2,000 foot breakwater began in 1867. It was completed in 1875 by the U.S. Government for less than its estimated cost of $240,000. Full credit for inception of the project has been attributed to the intelligent and energetic efforts of John Burt and his son Hiram.[71]

In 1866, John, his son Alvin, and William Burt joined several other investors, including Peter White and Samual P. Ely, to form the Marquette and Pacific Rolling Mill Company, located just south of Marquette. Its stack stood 52 feet high, nearly one-third taller than the prevalent charcoal furnaces. Other furnaces in the Upper Peninsula had relied on charcoal for fuel to produce cast pig iron, but this company's furnace became the first on Lake Superior to use bituminous coal.[72] The "Burt Mine" in Ishpeming supplied the hematite and slate ore to charge the furnace, and in 1872 the Marquette & Pacific Rolling Mill production reached 4,322 tons of pig iron and 622 tons of muck bar.[73]

John Burt House
Marquette's Pioneer Home - Marquette, Michigan   Built 1858 - Restored 1955

59.   John Burt house *courtesy Marquette County Historical Society*

The actual operation of the rolling mill began September 1, 1868, just three months after Marquette's great holocaust. On June 11, 1868, sparks from one of the Marquette & Ontonagon Railroad Company's locomotives touched off a blaze that quickly spread. Before it was over most of Marquette's business district and a good portion of the residential area was left in smoldering ruins.

John Burt, in Detroit at the time of the fire, returned to Marquette and began buying up the burned out properties, including the Jackson dock.[74] The Lake Superior Iron Company had used the Ely dock for its ore shipments until the fire completely destroyed it. John's display of confidence in Marquette led to further speculation that enabled the city to rapidly rebuild. In 1885 a *Mining Journal* article addressed John and his influence in this effort:

> When the morning's sun rose on the smoking and blackened ruins of every business house save one . . . the cry went forth from every lip, "Can Marquette survive this?" Many said no. There was needed a Moses to smite the rock of faith. Standing behind the firm of Burt Brothers, you gave that blow. . . ."[75]

As a result of the devastating fire Burt Brothers lost $158,000, two-thirds of which was uninsured.[76] The firm was the largest single financial loser. Within five years, however, all of their debts had been repaid with interest, and the "Burt block" was the first to be rebuilt.

60.   Burt Block in Marquette *courtesy of Marquette County Historical Society*

Later an ordinance was passed requiring all business establishments to be constructed of brick or stone. In response to the increased demand for these materials the Marquette Brownstone Company was formed in August 1872, followed two months later by the Burt Freestone Company. William Burt was one of the original incorporators in both sandstone companies, as well as the Huron Bay Slate and Iron Company, which was established to quarry and sell roofing slate.[77]

During the 1880's John Burt served as president of the Peninsular Iron Company as well as two additional companies located in Marquette: The Marquette Furnace Company and the Detroit & Marquette Iron Company. He also established the Union Fuel Company that built 11 kilns along the south bank of the Carp River.[78]

In addition to these ventures, John maintained an active role and financial interest in the Lake Superior Iron Company in Ishpeming that would yield more iron ore during the 19th century than any other iron company in Michigan.[79] Perhaps that is only fitting for the eldest son of the man who discovered Michigan's first great iron range.

In 1872 the Lake Superior Iron Company put into operation the Grace Furnace in Marquette, the first of its kind in Michigan designed to burn anthracite coal. Although it produced 30 tons of metal within the first week after the blast was started, and reached a high of 63 tons per day, the furnace was blown out in 1874 and production was never resumed.[80] Despite the many blast furnaces constructed in Marquette, there never was a major drive to turn the area into another Pittsburgh.

61.   Burt Brothers business card
        courtesy Marquette County Historical Society

John Burt - The Politician

Unlike his father who had been a Jacksonian Democrat, John Burt was an active member of the Republican party from its inception. He had considered himself a Democrat but changed his mind when, according to John, being a Democrat meant he had to assist in capturing runaway slaves crossing the Michigan border into Canada.[81] In 1862 John was chosen along with Philo Everett and J. W. Edwards to represent the Marquette County Republicans at the Houghton convention.[82]

In 1868 John was selected a Presidential elector at large by the Michigan Republican party.[83] On November 3, 1868, he was honored by the Electoral College with the opportunity to deliver to the president of the U.S. Senate Michigan's eight electoral votes for Ulysses S. Grant and his running mate Schuyler Colfax.[84] The following year the Marquette Republican Party organized the American Industrial League; John Burt was chosen its first president.

John Burt - The Inventor

John was an innovative civil engineer in addition to his many other activities. Between 1867 and 1884 he received a total of 12 patents from the governments of France, Canada, and the United States for his various inventions.[85] His patented creations included improvements in the manufacture of iron, a new method of purifying furnace gasses, a new system of ventilation, and a more economical process of obtaining charcoal for furnace use.

62.   Ore Dock in Marquette (1872) *Geological Survey of Michigan , 1873, I, 63.)*

65054—J. Burt—*Canal Lock.*

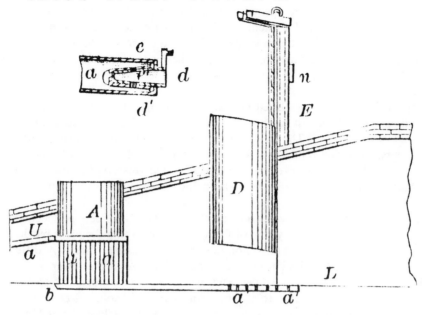

63.   John's lock patent drawing *(Patent Office records)*

64.   Weitzel Lock nearing completion in 1881. *courtesy MICHIGAN NATURAL RESOURCES MAGAZINE*

John's most ingenious idea, however, was his design for a canal-lock. While superintendent at the Soo Canal he saw the need for a new type of lock to overcome the turbulent wave of water that rushed against each vessel when the upper sluices were opened, which often caused serious damage and considerable delay. On May 28, 1867, John received a patent for his plan that eliminated the turbulence by admitting water through a series of ducts in the bottom of the lock and discharging the water through a series of pipes below the lower gate.[86] He estimated his plan could be incorporated into the existing lock at the Soo for about $75,000, thus saving the government the two million dollar cost of building an additional lock.[87]

In May 1870 General O. M. Poe was placed in charge of a project to prepare a plan for enlarging the Soo Canal and building a new lock just south of the existing one. Three months later his proposal was submitted to the Chief of Engineers.[88] The following year General Poe solicited suggestions from several Lake Superior engineers, including John Burt, for a new Soo Canal lock.[89] In February 1872 John sent General Poe the complete details of his patented plan. A modified version of it was then incorporated in the Weitzel lock that was completed in 1881, the year the Soo canal locks were transferred from state to federal control.[90]

The Weitzel lock was designed by General Godfrey Weitzel, General Poe's replacement in 1873. He denied any knowledge of Burt's patented system of filling and discharging the lock, but John's plan, sent to General Poe in 1872, was on file in Washington and easily accessible. The only difference between the Weitzel lock and John's patented design was the quantity and size of the lower ducts.

In 1882 John Burt initiated a $500,000 damage suit against the government for patent infringement. The case was unresolved when, on August 16, 1886, John died at his home in Detroit.[91]

Surprisingly, John Burt's role as a prime mover in the growth and development of Marquette has been largely ignored by historians. In his report on "The Influence of John Burt on Marquette and the Upper Peninsula," however, Alan D. Strocke concluded:

> [John] Burt's activities in the Upper Peninsula serve as a blue print to the growth of the area. . . . Burt did not merely venture into the area, speculate, and then wait for the developments to occur. Burt made those developments happen, and in this respect he must have enjoyed far greater satisfaction, than those who sought recognition for their contributions to the area.[92]

Chapter XXII

THE FINAL CHAPTER

*To be what we are, and to become what we
are capable of becoming, is the only end
of life.*[93]

- Robert Louis Stevenson (1850 - 1894)

The Burts vs. the U.S. Government

During the two decades from 1858 to 1877 the matter of William A. Burt's compensation for the federal government's use of the solar compass was not pursued by his heirs. John attributed this inactivity to the Civil War and "the unsettled condition of the country."[94] The case was renewed in December 1877, but the claim failed in the Senate due to insufficient documentation. New bills were introduced in both congressional chambers during the next eight years, but without success.

In late 1885, with advice from their lawyer not to abandon the fight, the heirs launched an all-out effort for approval of their claim. Concurrent bills were introduced in the House and Senate, each requesting a $250,000 appropriation. Signed petitions, letters, and other testimonials were solicited and received. During February 1886, John Burt coordinated the campaign from his Washington, D.C. hotel room. His son Hiram compiled the data and published a booklet that was distributed to members of Congress.[95]

One petition of endorsement from members of the Michigan Engineering Society led to a major controversy. John H. Mullett, son of the noted surveyor, submitted a paper to the engineering society, claiming his father had co-invented the solar compass and the Mullett heirs were equally entitled to any compensation granted to the Burts.[96]

Surveyor E. C. Martin suggested the possible motive for this action was that William A. Burt had uncovered a great deal of fraudulent survey work performed by James H. "Henry" Mullett. Martin added, "It may be that they have taken this way to get back at him."[97] During his lifetime John Mullett never claimed a role in developing the solar compass. In fact, he was present when William Austin assigned a half-interest in the instrument to Lucius Lyon. Contemporary surveyor-authors Ben Buckner and Carlisle Madson have concluded, "The literature on Burt's activities indicate he was the sole inventor."[98]

While in Washington during 1886, John Burt was also concerned with his own $500,000 lawsuit against the U.S. Government for infringement of his canal lock patent. At the time of John's death in August 1886, only months after he returned home from Washington, the Burts' claims against the government totaled $750,000.

Despite the strong campaign by William Austin's heirs, the Senate bill was defeated in March 1887. On November 29th of that year, Wells Burt died of "neuralgia of the heart," at age 67. During the 1890's much of the spark of the Burt heirs' lawsuit against the government was dampened with the deaths of William Austin's two remaining sons—Austin, at his home in Detroit, on February 18, 1894, at age 75, and William, in Marquette, on December 19, 1898, at age 73.

S. 3614.

IN THE SENATE OF THE UNITED STATES.

MARCH 15, 1900.

Mr. McMILLAN introduced the following bill; which was read twice and referred to the Committee on Public Lands.

A BILL

For the relief of the heirs of William A. Burt, deceased.

1   *Be it enacted by the Senate and House of Representa-*
2   *tives of the United States of America in Congress assembled,*
3   That the Secretary of the Treasury be, and hereby is, au-
4   thorized and directed to pay to the heirs of William A. Burt,
5   deceased, out of any money in the Treasury not otherwise
6   appropriated, the sum of two hundred and fifty thousand
7   dollars, as compensation for the use by the United States in
8   the survey of its public lands of the solar compass, the inven-
9   tion of said William A. Burt.

65.   Last bill for relief of Burt's heirs

John's son Hiram continued his father's patent infringement suit and his grandfather's solar compass claim, but he was unsuccessful in both cases. On March 15, 1900, with only Burt's grandsons left to continue the battle, the last bill for the relief of William Austin's heirs was introduced in the Senate.[99] It was referred to the Committee on Public Lands and died when the 56th Congress adjourned.

No further action has been taken to pursue the claim. Michigan historian Alan S. Brown has concluded, however, that:

> . . . to a generation many years removed from the nineteenth century, and able to view with more perspective and appreciation the services of William Austin Burt, it does not seem just that the Federal Government failed to reward the inventor in his lifetime.[100]

The Typewriter

After 1829, news of Burt's Typographer inspired several inventors to produce their own writing machines, while others worked independently without knowledge of the other devices. In 1852 John M. Jones of New York invented the "mechanical Typographer," using the term styled by Burt. Two years later R. S. Thomas of Wilmington, North Carolina, simply called his writing machine a "Typograph."

At least 13 additional typewriters were invented in the decade preceding 1868, the year Christopher Latham Sholes, Carlos Glidden, and S. W. Soule received their patent for a writing machine. Sholes, of Milwaukee, has been called "father of the typewriter" even though his first model was the 52nd, or possibly the 112th writing machine ever made.[101] Sholes had read about the details of a writing machine invented by American John Pratt, of Alabama, in the *Scientific American*, and he completed his own machine in September 1867.

Sholes is also credited with first using the term "typewriter," although the term was used in the *Scientific American* article detailing Pratt's machine.[102] After building nearly 50 different writing machines, Christopher Sholes finally produced one in 1873, which was called the "type-writer". Sholes then sold his rights to his financial backer, James Densmore, for about $6,000. Remington, the gunmakers, assumed the rights to manufacture the machine, but it was several years before a commercially successful product was marketed.

Today, more than not, William Austin Burt is given his due credit for inventing America's first typewriter. The author of a definitive biography of Sholes has acknowledged Burt's Typographer as the "first typewriter capable of practical work".[103] Earlier, an article in the U.S. Census Report for 1900 (Manufacturers) included a reference to Burt's invention as "the starting point of a great American industry."[104]

Burt's Equatorial Sextant

Matthew Fontaine Maury recommended the Equatorial Sextant to the U.S. Navy, but it was never purchased.[105] In 1857, however, when John Burt learned there was a need for an "artificial horizon" for use on ships at sea, he informed A. D. Bache of the Coast Survey:

> I have good reason to believe we can furnish such a horizonmone that will equal the best now known and used on landmwhich can be used on land, or at sea with even greater facility and accuracy than the present made.[106]

Bache replied that the cost of $50, suggested by John, would probably prevent it from being used extensively. He added, however, that "every public vessel ought to be provided with such as instrument if it fulfilled the requirement of accuracy and facility of use.[107]

In 1862 the classic Civil War confrontation between the Union's *Monitor* and the Confederate's *Merrimac* revolutionized naval warfare. The following year the Navy Department established a committee to develop an instrument to accurately determine the variation of the needle on naval vessels. When William J. Young learned of this action he submitted his own plan for such an instrument to A. D. Bache. Young would alter Burt's Equatorial Sextant to include only a latitude arc, an hour arc, and an azimuth circle.[108]

Bache declined Young's proposal, but Young persisted, indicating that "a part of Burt's Equatorial Sextant will be the thing needed, and in every way better than any simplification or modification of the solar compass."[109] Bache agreed to discuss the matter with Young on his next visit to Philadelphia, but, except for experimental models, the instrument was never made.[110]

In 1940, as Hitler's Nazi troops occupied France, a Burt descendant attempted to interest the U.S. Navy Department in adopting the Equatorial Sextant for quick and accurate computation of nautical and aerial navigational problems.[111] He was unsuccessful.

The Solar Compass

Although William A. Burt and his heirs were never compensated for the use of the solar compass in the government surveys, the value of the instrument has been fully acknowledged. In his annual report for 1875 the General Land Office Commissioner reported that "the chain and solar compass are the principal instruments of execution [in the public land surveys]." From information provided by the land commissioner, John Burt estimated that the government had saved nearly $15,000,000 by using the solar compass in the land surveys between the years 1849 and 1878.[112]

The solar compass was used in conjunction with a portable astronomical transit on the California-Nevada boundary survey in 1863.[113] The technique, according to California surveyor-author Bud Uzes, was to use the astronomical transit to extend the main line across the mountains from peak to peak, while the solar compass was used between transit stations on the peaks to actually run the line on the ground.[114]

In the late 1860's Butler Ives, brother of surveyor William Ives, utilized the solar compass during his preliminary survey of the route for the Central Pacific Railroad, between Lake Tahoe in the Sierra Nevada Mountains and Utah's Great Salt Lake. He also used the solar compass to explore the route between Salt Lake and Green River, Wyoming, currently a part of the Flaming Gorge Recreation Area. Following this work, Ives wrote:

> The solar compass has proved a very valuable instrument in these explorations. With it I have made in the last two years, more explorations of the country with one small party, than could have been done with two parties with most any other instrument, and at much less cost than was expected when I started out.[115]

The solar compass, used by William A. Burt to locate the rich Marquette iron range in 1844 also enabled Edwin J. Hulbert to discover Michigan's most productive copper lode, the Calumet and Hecla Mines, in 1864. As John Burt explained:

> While copper does not attract the magnet, the Solar Compass first discovered . . . that trap rocks did attract the needle, and that copper veins, in conjunction with sedimentary and trap rocks, also attracted the magnet, probably from galvanic currents along the line of junction.[116]

In 1886 Michigan's Commissioner of Mineral Statistics, Charles D. Lawton, told John Burt that he had personally used Burt's Solar Compass for about 25 years. Lawton added:

> It is impossible to conceive how the public lands could have been run out with any degree of accuracy or rapidity without the aid of this instrument. . . . Taken all in all, I regard the Solar Compass as one of the most useful inventions ever made in this country.[117]

66.   Surveyor's chain (1875) *courtesy of Cecil Hanson (from Keuffel & Esser Engineers' and Surveyors' Illustrated Catalog.)*

67.   1878 Burt solar compass made by W. & L.E. Gurley for the U.S. Geological Survey *courtesy Smithsonian Institution*

151

By 1894, when use of the magnetic compass was prohibited on the U.S. public surveys, the solar transit was in common use. In 1867 William Schmolz of San Francisco patented an instrument that mounted a portion of the Burt solar mechanism as an attachment onto an engineer's transit. Between 1874 and 1884 W. & L.E. Gurley Company paid him $5.00 royalty for each unit that was sold.[118] The adaptation, however, did not add to the accuracy of the solar compass, which operated with greater speed.[119] In 1880 surveyor Benjamin H. Smith developed a telescopic solar attachment that was combined with the engineer's transit. Although many different forms of attachment were considered, in 1914 the unit developed by Smith became the prototype for the modern BLM solar transit.[120]

In 1920 a GLO official reported:

> There has been a continuous use of the Solar Compass in the survey of the public lands since its invention to the present day and there is no prospect of its discontinuance. . . . [The solar transit] cannot be said to have replaced the solar compass; rather it may be said that each instrument has had its own field of action and the two have in many cases been used in the same party, each for the particular purpose to which it was best adapted."[121]

Today, in the 1980's, the BLM office in Portland still maintains a supply of Burt solar compasses for special situations. According to William W. Glenn, Chief, Cadastral Survey, including:

- field investigations and reconnaissance surveys where small amounts of line are run but true bearings are desired.

- corner remonumentation projects because a true bearing can be easily determined for use in locating the corner point from corner accessories.

- as an aid in training new surveyors the practices and equipment used in older government surveys.[122]

William Austin Burt would have been justly proud of his lasting contribution.

The Marquette Iron Range

W. S. Bayley, in his report for the 1897 U.S. Geological Survey, accorded William Austin his full credit in locating the Marquette iron range:

> It is to Burt's energy and to his discovery of ore that later developments of the iron district are due."[123]

Significant production of iron from the Marquette range began in 1854. Since 1910, 81 percent of all the iron ore shipped in the United States has come from the Lake Superior region.[124] The value of the iron and copper shipped from the Great Lakes area has exceeded that of all the gold ever mined in California.[125] In 1895 iron ore production from the Mesabi range in Minnesota surpassed that of the Marquette range, and it has never relinquished its leading position.[126]

Evidence of the Marks

William A. Burt has been honored in Michigan with a lake, a state park, a township, and a mountain perpetuating his name. Burt Lake, called "one of the prettiest lakes ever looked upon by the tourist," is located in Cheboygan County, about 25 miles south of the magnificent Mackinac Bridge that connects

Michigan's Upper and Lower Peninsulas.[127] It is one of Michigan's largest inland lakes, about nine miles long and five miles wide. Today it is a popular recreational area—a part of the 405 acre Burt Lake State Park.[128]

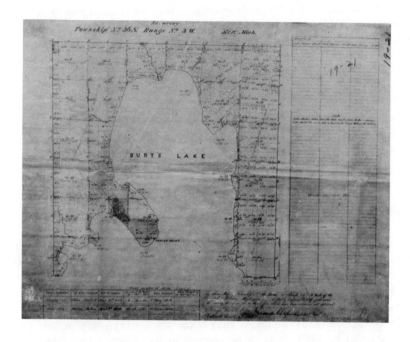

68.   Burt Lake plat map - 1841 *courtesy Michigan State Archives*

Burt surveyed in the area during early 1840 when he carried up a line between Ranges 3 and 4 West, past the western edge of the lake to the Straits of Mackinaw. His work enabled John Mullett to run the initial township lines during the second quarter of 1840.[129] Mullett's original plat map, dated February 24, 1841, named the body of water "Burt's Lake," for the first time. It is fitting that Mullett Lake is located nearby, just east of Burt Lake.

Burt Township, embracing 16 sections in the area around Burt Lake, was first organized on February 22, 1860, with its post office officially established the following October.[130] This recognition of William Austin Burt, however, proved ephemeral. During the next 40 years portions of Burt township were either annexed to other townships, or became new townships within the county. The last portion of Burt Township in Cheboygan County, Michigan was annexed in June 1916.

There were actually two Burt Mountains in Michigan's Upper Peninsula. One, owned by the Lake Superior Iron Company, or "Burt Mine", was located 17 miles from Lake Superior, one mile west of the Cleveland Mountain.[131] It was named for John Burt and his brothers who had invested in the mine. The other, first named "Mount Burt" by geologist John Locke in 1847, is located in the Huron Mountains.[132] For several years it honored William Austin Burt; subsequently it was renamed "Sugarloaf Mountain".[133]

Today, nearly 140 years after the Burts marked the survey lines in Michigan and the midwest, evidence of their work is still being uncovered. Retracement surveyor Vic Hedman of the U.S. Forest Service recently acknowledged:

> We who have and are retracing the surveys of W. A. Burt in Upper Michigan are very much impressed by the accuracy of Mr. Burt's surveys. We perpetuate the recovery of the corner evidence established by Mr. Burt with brass capped iron pipes.[134]

Surveyor Donald D. Lappala with Upper Michigan's Ottawa National Forest Service also wrote:

> [William A.] Burt, known for his solar compass, was an outstanding land surveyor and an ingenious inventor. A tribute to Mr. Burt is the accuracy with which he established evidence in accordance with the record of his surveys. This accuracy has provided a strong linkage and proof for the recovery of land corner evidence.[135]

Sadly, Lappala added:

> It is reasonable to estimate that all remaining bearing tree evidence of the corners of the original public land survey on the upper peninsula of Michigan will be gone by the end of this century.

Other Recognition

Between 1920 and 1955 William A. Burt was nominated six times to The Hall of Fame for Great Americans.[136] Although the Michigan Historical Commission and the University of Michigan jointly acted upon a resolution of the state legislature of 1929 to present Burt as a Hall of Fame candidate, a sufficient number of votes were not received.[137] Through 1977, when new elections were postponed indefinitely, only nine U.S. inventors—including the Wright brothers—had been chosen.[138]

To honor the site of the original William A. Burt homestead, which he purchased in 1822, a historical marker was commissioned in 1985. It is to be erected in the Stony Creek Metropark, in Washington township, Macomb County, on property originally belonging to Burt.[139]

Reference Notes

I. In The Beginning:

[1] Franklin Pratt, "The Taunton Burts," *The Life and Times of Henry Burt of Springfield* (1893), pp.499–530.

[2] *Taunton Architecture: A Reflection of the City's History* (Taunton, Mass, 1981), p. 7.; John W. Barber, Historical *Collections Relating to the History and Antiquities of Every Town in Massachusetts* (1839), pp.141–43; Pratt, p. 500.

[3] Pratt, p. 513.

[4] Federal Writers Project, *Massachusetts: A Guide to its Places and People* (Boston: Houghton Mifflin, 1937), pp. 367–70.

[5] William A. Burt, "Autobiography," *Michigan Pioneer Historical Collections* (40 volumes, Lansing, 1877–1929), XXVIII, 646–47; hereafter cited as *MPHC*.

Sally Burt (1797-1877) married Jonathan Searles on May 25, 1814. Their son Zelotes Searles, born in 1815, later learned the art of surveying with the solar compass from his uncle William A. Burt. Searles was a "good and conscientious surveyor" and "he did considerable surveying for the U.S. Government in the mineral region on the south shore of Lake Superior." (William Burt, Letter to Franklin Pratt, Dec. 19, 1894, Burt Papers, Taunton Historical Museum).

[6] George H. Cannon, "The Life and Times of William A. Burt of Mt. Vernon, Michigan," *MPHC*, V, 115–23.

[7] Alvin Burt (1780-1853), the eldest son of Alvin and Wealthy, did become a sea captain.

[8] Horace Eldon Burt, "William Austin Burt, Inventor," *Michigan History Magazine*, (1922), VI, 176.

[9] Genealogical notebook compiled by Austin Burt, great grandson of William Austin Burt, author's collection.

[10] Austin Burt genealogical notebook.

[11] Wm. A. Burt, "Autobiography."

[12] Cannon, pp. 115–23. It is unknown if Burt's diary still exists; however, it is frequently cited by Cannon.

[13] Alec R. Gilpin, *Territory of Michigan Eighteen Hundred FivemEighteen Hundred Thirty-Seven* (Mich. St. Univ. Press, 1970), pp. 128–30.

[14] Tiffin's letter is reprinted in Burton H. Boyum (ed), *The Mather Mine* (Marquette County Historical Society, 1979), p. 14.

[15] Gilpin, p. 131–32.

[16] W.B. Williams, "The Surveyor, Pioneer to Professional," *Surveying and Mapping,* X (Oct.–Dec., 1950), 256–61.

[17] Hervey Parke, "Reminiscences," *MPHC*, III, 572–87; Walter Havighurst, *Wilderness For Sale* (N.Y.: Hastings House, 1956), pp. 84–86. Hervey Parke (1790–1879) was named after his uncle Hervey Parke, a co-founder of the Parke-Davis Pharmaceutical Company.

[18] George H. White, "A Sketch of Lucius Lyon, One Of The First Senators From Michigan," *MPHC*, XIII, 322–33; J. M. Schmiedeke, "With Chain and Compass," *Surveying and Mapping* (March, 1972), pp. 63–67.

II. Opportunity in Michigan Territory:

[19] Parke, p. 577.

[20] Austin Burt ,"Reminiscences of Wm. Austin Burt, Inventor of the Solar Compass," *Michigan History Magazine,* VII (Jan.–Apr., 1923), 32–41.

[21] Austin Burt, p. 36.

[22] Austin Burt[10] Papers, author's collection.

[23] Gilpin, p. 25.

[24] Clarence E. Carter (ed.), *The Territorial Papers of the United States* (Washington: 1943), XI, 1036–37. Burt and Stockton knew each other since the War of 1812, when both served in the New York militia.

[25] Gilpin, p. 108; *Journal of the Legislative Council of the Territory of Michigan—Second Session of the Second Council* (1827), index.

[26] John Mullett, Letter to Samuel Williams, Chief Clerk, Sur. Genl. Office, Cincinnati, Ohio, April 6, 1827, Williams Papers, William L. Clements Library, Univ. of Michigan; hereafter cited as Williams Papers.

[27] Edward Tiffin, Letter to George Graham, Comm., GLO, Feb. 11, 1826, cited in Carter, XI, 949–52.

[28] Parke, p. 581.

[29] Cannon, pp. 115–23.

[30] Charles Fey, "William Austin Burt Surveyor of the Upper Peninsula," *The Masonic World* (June 1945), p. 16.

[31] Cannon, pp. 115.23.

[32] Fey, p. 16.

III. America's First Typewriter:

[33] *The Republic,* Book II, Chapter XI, p. 369.

[34] Horace E. Burt, *William Austin Burt, Inventor* (Chicago, 1920), p. 14, n. 1.

[35] "Jackson and the Michigan Judiciary," *Michigan History Magazine,* LIX (Fall, 1975), 138–143; Silas Farmer, The History of Detroit and Michigan, 2nd ed. (Michigan: Silas Farmer & Co., 1889), I, 671–672.

[36] Horace Eldon Burt, *William Austin Burt, Inventor,* 175–93; Wilfred A. Beeching, *Century of the Typewriter* (New York: St. Martin's Press, 1974), pp. 4–7; Bruce Bliven, Jr., *The Wonderful Writing Machine* (New York: Random House, 1954), pp. 24–41.

[37] Several authors have inaccurately written that the Typographer contained a roll of paper. Specifications in the letters patent (National Archives Record Group No. 241), however, show that a single sheet was attached to a rotating belt. The handwritten notes of Austin Burt, who built a reproduction of the Typographer for the Smithsonian Institution, in 1893, support this. (Author's collection).

[38] Gilpin, p. 95.

[39] Bliven, p. 31; Arthur Toye Foulke, *Mr. Typewriter: A Biography of Christopher Latham Sholes* (Boston: Christopher, 1975), p. 52.

[40] Letters Patent specifications and drawing (restored in 1893 from a parchment copy of the original patent, obtained from Burt family papers), National Archives Record Group 241. See "An Historical Letter," *Office Appliances* (Chicago, June, 1922)

[41] "The First Typewriter," *The Detroit Free Press,* Feb. 5, 1888, Burt Papers, Marquette County Historical Society, Marquette, Mi., hereafter cited as Burt Papers.

[42] John P. Sheldon, Letter to Eliza Sheldon, Oct. 29, 1829, Sheldon Papers, Burton Hist. Coll., Detroit Public Library.

[43] Bliven, p. 32. The original specifications do not include any reference to legs supporting the typographer frame. The patent office drawing (1893 restoration) includes them, however.

[44] Horace Eldon Burt, *William Austin Burt, Inventor,* pp. 175–93. Austin Burt's reproduction of the Typographer and the March 13, 1830, letter from Burt to Phebe were donated by the Burt family to the Smithsonian Institution's National Museum of American History in 1923. A second reproduction was built in Washington, D.C., in 1924, for the Science Museum, South Kensington, London.

[45] Wm. A. Burt, Letter to Cyrus Spalding, March 17, 1830, Burt Papers.

[46] Cited in "The First Typewriter", *The Detroit Free Press,* February, 5, 1888., Burt papers.

[47] Cited in Rupert T. Gould, *The Story of the Typewriter* (London: Gee and Company Limited, 1949), pp. 11–15.

[48] Bliven, p. 25.

[49] Beeching, p. 4.

[50] Michael H. Adler, *The Writing Machine* (London: George Allen & Unwin, Ltd., 1973), pp. 365–66.

IV. A New Career:

[51] Walter Romig, *Michigan Place Names* (Romig, n.d.), p. 383.

[52] Austin Burt, "Reminiscences of Wm. Austin Burt, Inventor of the Solar Compass," *Michigan History Magazine,* VII (Jan.– Apr., 1923), 32–41.

[53] George W. Thayer, "Life of Lucius Lyon," *MPHC* (1896), XXVII, 406; J. M. Schmiedeke, "With Chain and Compass," Surveying and Mapping, March, 1972, pp. 63–67.

[54] Cited in Lucuis Lyon, Letter to M. T. Williams, Surv. Genl., Aug. 31, 1833, Burt Papers.

[55] Lucuis Lyon, Letter to M. T. Williams, Surv. Genl., Aug. 31, 1833, Burt Papers. Lyon sent William A. Burt a copy of this letter on Sept. 1, 1833.

[56] Lowell O. Stewart, *Public Land Surveys, History, Instructions, and Methods* (Ames, Iowa, 1935), p. 6.

[57] Cited in C. Albert White, *A History of the Rectangular Survey system*, U.S. Dept. of the Interior, Bureau of Land Management (Washington, D.C.: GPO, 1984), pp. 11–15.

[58] Stewart, p. 35.

[59] *American State Papers,* Public Lands (1789–1809), I, 73; cited in Stewart, pp. 35–36.

[60] Austin Burt, pp. 38–39; John Bartlow Martin, *Call it North Country, The Story of Upper Michigan* (New York: Alfred A. Knopp, 1944), Ch. 4, "Chain," pp.50–51.; Knox Jamison, "The Survey of the Public Lands in Michigan," *Michigan History Magazine,* XXXXII, (June, 1958), 197–214.

[61] Burt's instructions and field notes are located in the State Archives, Michigan Department of State, Lansing.

[62] Austin Burt, p.38.

[63] Martin, 51–52.

[64] Jamison, pp. 199–200.

[65] Hervey Parke, Letter to Micajah T. Williams, Surv. Genl., Feb. 3, 1834, GLO-SGO, Ltrs.Recd., XVIII, 1834, cited in Carter, XII, 722.

V. Quest For Accuracy:

[66] John Wilson, Commissioner, Genl. Land Office, Letter to Sen. Alpheus Felch, Chairman, Comm. on Public Lands, Mar. 4, 1852, cited in Hiram A. Burt, *Facts Upon Which is Based the Claim of the Heirs of Wm. A. Burt, Deceased* (Wash., D.C.: 1886), p. 9.

[67] Lucius Lyon, Letter to Wm. A. Burt, Oct. 25, 1834, Burt Papers; cited in Alan S. Brown, "William Austin Burt: Michigan's Master Surveyor," *Papers of the Michigan Academy of Science, Arts, and Letters,* (1962), XIVII, 263–75.

[68] Austin Burt, pp. 38–39.

[69] George H. Cannon, "William Austin Burt: The Tale of a Pioneer," *Mount Clemens Monitor* (n.d.), Burt Papers, Burton Historical Collection, Detroit Public Library.

[70] George H. Cannon, "The Life and Times of William A. Burt," *MPHC* (1882), V, 115–23; Austin Burt, pp. 32–41. In 1834 John Burt purchased a large tract of land near Homer Village, located south of Albion near the St Joseph River. (History of Calhoun County, Michigan (1877), p. 121.)

[71] M.T. Williams, Surv. Genl., Letter to Wm. A. Burt, Dec. 27, 1834, *Surveyor General's Letter Book,* Iowa Dept. of State. Burt's contract dated December 23, 1834 was cancelled and another was drawn up to include a strip of four fractional townships along Lake Michigan, extending northward from the center of modern day Milwaukee.

[72] Alice E. Smith, *The History of Wisconsin* (Historical Society of Wisconsin, 1973), I, p. 195.

[73] Hiram Burnham, Letter to Micajah T. Williams, Feb. 16, 1835, GLO:SGO Ltrs. Recd., XX, 1835:ALS; cited in Carter, XII, 862.

[74] Hervey Parke, Letter to Micajah T. Williams, Feb. 17, 1835, GLO-SGO, Ltrs. Recd., XX, 1835; cited in Carter, XII, 863.

[75] White, pp. 82–83.

[76] Hiram Burnham, Letter to Micajah T. Williams, July 11, 1834, GLO-SGO Ltrs. Recd. XIX, 1834:ALS; cited in Carter, XII, 786–88.

[77] Lucius Lyon, Letter to Micajah T. Williams, March 28, 1832, GLO:SGO, Ltrs. Recd., XV, 1832: ALS

[78] John Mullett, Letter to Micajah T. Williams, Apr 20, 1832, GLO:SGO Ltrs. Recd., XV, 1832: ALS, cited in Carter, XII, 467–68.

[79] Adin Baber, *Lincoln with Compass & Chain* (A.H. Clark, 1968), p. 35.

[80] *General Instructions to His Deputies by the Surveyor General for Ohio, Indiana, and Michigan Territory* (1833); cited in White, p. 293.

[81] Hiram Burnham, Letter to Micajah T. Williams, April 24, 1835, GLO:SGO, Ltrs. Recd., XX, 1835, ALS; cited in Carter, XII, 900–01.

[82] Wm. A. Burt, Letter to Micajah T. Williams, May 17, 1835, GLO:SGO, Ltrs. Recd., XX, 1835:ALS; cited in Carter, XII, 921–22.

[83] William A. Burt, Letter to Phebe Burt, May 17, 1835, Burt Papers; cited in Silas Farmer, Biographical ed., p. 1180.

[84] Silvio A. Bedini, *Thinkers and Tinkers, Early American Men of Science,* (Rancho Cordova, CA: Landmark, 1983), p. 474.

[85] Thomas Freeman, Surv. Genl., Letter to Silas Dinsmore, *Instructions to the Principal Deputy Surveyor of the Land District East of Island of N. Orleans* (about 1815); cited in White, pp. 251–54.

[86] Burton Boyum, "The Compass That Changed Surveying," *Professional Surveyor,* Sept./Oct. 1982, pp. 28–31.

[87] The British Nautical Almanac, which contained tables of astronomical data, was first introduced in 1767. The United States Nautical Almanac has been published since 1855.

[88] The "co-latitude" was actually used.

[89] Francois D. Uzes, "Operating Principles of a Burt's Solar Compass," *The California Surveyor* (Summer 1981), pp. 14–15.

[90] The Mining Journal (Marquette), March 2, 1878.

[91] Bedini, pp. 370–71. According to Bedini, the invention of the surveyor's transit has also been attributed to Edmund Draper, a Philadelphia instrument maker of the same period; Charles E. Smart, *The Makers of Surveying Instruments in America Since 1700* (New York: Regal Art Press, 1962), pp. 172–74.

[92] Bruce Sinclair, *Philadelphia's Philosopher Mechanics: A History of the Franklin Institute, 1824-1865* (Baltimore: John's Hopkins, 1975), p. x.

[93] Bache later established the first magnetic observatory in North America, headed the U.S. Coast Survey, and he became founder and first president of the National Academy of Sciences.

[94] Other recipients of the John Scott Legacy Medal include George Westinghouse, Jr. (1874); Carlos Gliddens, C. Latham Sholes & Samuel W. Soule (1875); Pratt & Whitney Co. (1883, 1886); and Thomas A. Edison (1889) The last Scott's medal was awarded in 1918.

[95] William J. Young, Letter to Wm. A. Burt, Jan. 9, 1836, Burt Papers.

[96] Austin Burt, "Burt's Solar Compass," *MPHC*, (1886), XXXVIII, 191.

[97] Wm. A. Burt, Assignment to Lucius Lyon, Sept. 10, 1836, Lyon Papers, William L. Clements Library, Univ. of Michigan, hereafter cited as Lyon Papers. This assignment was later officially recorded by Lyon at the U.S. Patent Office, on July 16, 1841. (see Lucius Lyon to H.L. Ellsworth, Comm. of Patents, July 10, 1841, "Letters of Lucius Lyon, One of the First United States Senators From Michigan," *MPHC*, (1896), XXVII, 412-601.

VI. Trial and Error:

[98] Translated from *The Prince*, Ch. 26.

[99] E. A. Brown, Letter to R. T. Lytle, Aug. 16, 1836; cited in Roscoe L. Lokken, *Iowa Public Land Disposal* (State Hist. Soc.of Iowa, 1942), pp. 105–06.

[100] cited in Lokken, p. 18.

[100] Wm. A. Burt, Letter to Robert Lytle, Dec. 8, 1836, *Surveyor General Letter Book:* 1832–1840, File A, State Land Office, Des Moines, Iowa.

[101] Stewart, p. 114.

[102] Hervey Parke, "Reminiscences," pp. 588–99.

[103] E. R. Harlan, "Original Field Notes of William Austin Burt of the Survey of the Fifth Meridian (now in) Iowa, November, 1836," *Annals of Iowa*, Oct. 1935, (3), XX, 83–112.

[104] Harlan, p.83.

[105] Wm. A. Burt, Letter to Lucius Lyon, April 6, 1837, Lyon Papers.

[106] Wm A. Burt, Letter to Morgan L. Martin, July 27, 1837, Martin Papers, box 5, Wisc. State Historical Society Archives.

[107] Bayrd Still, *Milwaukee, The History of a City* (Madison: Hist. Soc. of Wisconsin, 1948), pp. 14–15.

[108] Wm. A. Burt, Letter to Wm. Hamilton, Aug. 11, 1837, Lyon Papers.

[109] Wm. J. Young, Letter to Wm. A. Burt, Aug. 3, 1838, Burt Papers.

[110] Wm. J. Young, Letter to Wm. A. Burt, Oct. 23, 1838, Burt Papers.

[111] Wm. J. Young, Letter to Wm. A. Burt, Apr. 27, 1839, Burt Papers.

[112] Wm. J. Young, Letter to Wm. A. Burt, Sept. 27, 1839, Burt Papers.

[113] E. S. Haines, Surv. Genl., Letter to Wm. A. Burt, Aug. 15, 1838, *Surveyor General Letter Book*, Iowa Dept. of State, p. 281.

[114] Cited in Stewart, pp. 88–89.

[115] Estimate compiled by Robert C. Miller, Pittsburgh, Pa., authority on Wm. J. Young, from letters of correspondence between Young and Wm. A. Burt, Burt Papers. Young gave credit to his workman, Mr. Ford, for building the first model of Burt's variation compass, in 1835. A Mr. Timson, of Toronto, in 1887, claimed to have made the first solar compass for Burt. (Annual Report of the Provincial Land Surveyors, 2nd Annual Meeting, 1887, p. 13.) While this is unsubstantiated, Timson may have been referring to the term applied to Burt's compass about 1840.

[116] Smart, p. 162.

[117] Wm. A. Burt's petition to Congress, Jan. 4, 1850, cited in Hiram Burt, *Facts,* pp. 4–5. Burt later indicated that he usually waived the royalty fee so more surveyors would be encouraged to purchase the solar compass.

[118] Wm. J. Young, Letter to Wm. A. Burt, Apr. 18, 1841, Burt Papers.

[119] R. Whitcomb, Letter to Wm. A. Burt, July 19, 1843, Burt Papers.

[120] R. Whitcomb, Letter to Wm. A. Burt, Mar. 14, 1848, Burt Papers.

[121] Cannon, "The Life and Times of William A. Burt," p. 118.

VII. Acquiring the Upper Peninsula:

[1] George W. Thayer, "Life of Senator Lucius Lyon," *Michigan Pioneer and Historical Collections* (hereafter cited as MPHC), XXVII (1896), 405–11. At the Democratic territorial convention, in February, 1833, Lyon defeated John P. Sheldon and Austin E. Wing.

[2] Willis F. Dunbar, Michigan: *A History of the Wolverine State*, (Erdmans, 1965), pp. 308–11.

[3] L. G. Stuart, "Verdict For Michigan. How the Upper Peninsula Became a Part of Michigan.," *MPHC* (1896), XXVII, 390–403.

[4] U.S. Cong., House, Robert Lytle, Surv. Genl., *Report to James Whitcomb, Comm. GLO*, Dec. 18, 1837, 25-2, U.S. Serial 327, H. Doc. 197, pp. 2–3.

[5] John H. Forster, "Reminiscences of the Survey of the Northwestern Lakes," *MPHC* (1886), IX, 102.

[6] *History of Saginaw County, Michigan* (1881), p. 291.

[7] Cited in Richard Reeves, *American Journey, Traveling with Tocqueville in Search of Democracy in America* (New York: Simon and Schuster, 1982), p. 177.

[8] Cited in *The Public Lands, A Brief Sketch in United States History*, U.S. Dept. of The Interior, BLM, (1963), p. 1.

[9] U.S. Cong., House, Robert Lytle, Surv. Genl., *Report to James Whitcomb, Comm. GLO*, Dec. 18, 1837, 25-2, U.S. Serial 327, H. Doc. 197, pp. 2–3.

[10] U.S. Cong., House, Lewis Clason, Letter to R. T. Lytle, Surv. Genl., Aug. 13, 1837, 25-2, U.S. Serial 327, H. Doc. 197, pp. 6–7.

[11] Contracts dated Aug. 25, 1838.

[12] A. G. Ellis, Surv. Genl., Instructions to Wm. A. Burt, Oct. 24, 1839, Surveyor *Generals Letter Book*, State Land Office, Des Moines, Iowa.

VIII. The Upper Peninsula Surveys Begin:

[13] From Spencer's *Social Tactics,* 1850.

[14] E. S. Haines, Surv. Genl., Special Instructions to Wm. A. Burt, D.S. for Running Exterior Lines in Michigan, Jan. 14, 1840.

[15] *Annual Report of the State Geologist,* Feb. 3, 1840, cited in George N. Fuller (ed), *Geological Reports of Douglass Houghton, First State Geologist of Michigan, 1837–1845* (Lansing: Michigan Hist. Commission, 1928), pp. 367–98.

[16] Wm. A. Burt, Letter to Phebe Burt, June 24, 1840, Burt Papers, Marquette County Historical Society, Marquette, Michigan; cited in Alan S. Brown, "William Austin Burt: Michigan's Master Surveyor," *Papers of the Michigan Academy of Science, Arts, and Letters* (1962), XLVII, 269.

[17] Ernest H. Rankin, "Burt Monument Gives Point of Original Survey In State," *The Mining Journal,* Oct. 26, 1964.

[18] Cited in William Austin Burt, *Description of the Solar Compass* (Detroit: Geiger & Christian, 1844), pp. 25–26.

[19] Wm. A. Burt, Letter to Alvin Burt, Feb. 17, 1841, Burt Papers.

[20] Cited in William Austin Burt, *Description of the Solar Compass,* p. 26.

[21] *Niles National Register,* May 1, 1841, LX, 131; also cited in Stewart, pp. 88–89.

[22] C. Albert White, *A History of the Rectangular Survey System,* U.S. Dept. of the Interior, BLM (Wash., D.C.: GPO, 1984), p. 99.

[23] George Cannon, "The Life and Times of William A. Burt of Mt. Vernon, Michigan," *MPHC,* V, 115–23.

[24] Wm. A. Burt, Letter to Alvin Burt, Feb. 17, 1841, Burt Papers.

[25] Wm. J. Young, Letter to Wm. A. Burt, Apr. 18, 1841, Burt Papers.

[26] John H. Mullett, "Invention of the Solar Compass," *Proceedings of the Michigan Engineering Society* (1888), p. 100.

IX. Extending the Upper Peninsula Surveys:

[27] Wm. A. Burt, Letter to Wm. Woodbridge, U.S. Senator from Michigan, Jan. 3, 1842, Woodbridge Papers, Burton Hist. Collection, Detroit Public Library, Detroit, Mich. Hereafter cited as Woodbridge Papers.

[28] Chas. Rodd, Receipt (n.d.) to Wm. A. Burt, Burt Papers

[29] Wm. A. Burt, Letter to Alvin Burt, Feb. 17, 1841, Burt Papers.

[30] White, p. 99.

[31] Wm. A. Burt, Letter to Ezekiel S. Haines, April 16, 1841, Burt Papers.

[32] Lucius Lyon, Letter to H. L. Ellsworth, Comm. of Patents, July 10, 1841, "Letters of Lucius Lyon, One of the First United States Senators From Michigan," MPHC (1896), XXVII, 412–601.

[33] Thayer, p. 408.

[34] Fuller, pp. 9–13.

[35] T.B. Brooks, *Geological Survey of Michigan: Upper Peninsula, 1869–1873* (New York: Bien, 1873), I, 11.

[36] Ferris E. Lewis, *Michigan Since 1815* (Hillsdale, 1973), p. 71.

[37] Douglass Houghton, Letter to Wm. Woodbridge, U.S. Senator, Dec. 16, 1840, cited in George N. Fuller (ed), *Geological Reports of Douglass Houghton, First State Geologist of Michigan, 1837–1845* (Lansing: Michigan Hist. Commission, 1928), pp. 473–81.

[38] Douglass Houghton, Letter to Wm. Woodbridge, Dec. 16, 1840.

[39] Dr. Houghton also added the following footnote: "These remarks are intended to apply directly to the trap region. Beds of bog iron occur, east from Chocolate river, which probably may at some future date be profitably worked." The major iron ranges of Michigan, first located by Burt in 1844, were at least seven miles inland from Lake Superior and several miles west of the Chocolate River.

[40] Houghton's Geological Report of 1841, Fuller, p. 527; also cited in Brooks, p. 11.

[41] Lew Allen Chase, "Early Days of Michigan Mining: Pioneering Land Sales and Surveys," *Michigan History Magazine,* XXIX, 175.

[42] E. S. Haines, Special Instructions for running town lines in Michigan, to Wm. A. Burt, April 2, 1841.

[43] White, p. 100.

[44] Cited in Fuller, p. 668. In 1842, Dr. Houghton served as Mayor of Detroit.

[45] White, p. 100.

[46] William A. Burt, D.S., "Geology, Topography, & Meteorology of Upper Michigan, 1841", handwritten notebook, Bentley Library, Univ. of Michigan.

[47] Wm. A. Burt to William Woodbridge, U.S. Senator, Jan. 3, 1842, Woodbridge Papers; also see Emily George R.S.M., *William Woodbridge, Michigan's Connecticut Yankee* (Michigan History Div., Michigan Dept. of State, 1979), p. 59. The U.S. Government was apparently seeking to borrow 11 million dollars. Burt requested a minimum return of 6% on his investment.

[48] E. S. Haines, Surv. Genl., Instructions to Wm. A. Burt, April 11, 1842, Michigan Dept. of State, State Archives.

[49] William A. Burt, field notes sent to E. S. Haines, May 2, 1842. Original field notes located in Michigan Department of State, State Archives.

[50] White, p. 100.

[51] Brooks, p. 12.

[52] U.S. Cong., House, Wm. Johnson, Surv. Genl., *Annual Report of 1843*, to Thomas Blake, Comm., GLO, Oct. 1, 1843, 28-1, U.S. Serial 441, H. Doc. 37, pp. 52–55.

X. A Mountain of Iron:

[53] William A. Burt, *Description of the Solar Compass Together with Directions for its Adjustment and Use* (Detroit: Geiger & Christian, Printers, 1844), 27 pages.

[54] For a profile of Rasselas P. Whitcomb (1808–1884) see Charles E. Smart, *The Markers of Surveying Instruments in America Since 1700* (Regal Art Press, 1962), p. 258. A "Burt Patent" solar compass, made by R. Whitcomb, is frequently exhibited by the Surveyors Historical Society (California).

[55] Silvio A. Bedini, *Thinkers and Tinkers, Early American Men of Science* (Rancho Cordova, CA: Landmark, 1983), pp. 361–69; Roger Gerry, "Richard Patten: mathematical instrument maker," *Antiques* (July 1959), pp. 56–58.

[56] Bedini, pp. 361–62.

[57] Lowell O. Stewart, *Public Land Surveys, History, Instructions, and Methods* (Iowa: Meyers Printing, 1935), p. 89.

[58] Wm. A. Burt, Letter to Lucius Lyon, Feb. 29, 1844, Lyon Papers, Clements Library, Univ. of Michigan.

[59] Solon Burt (son of Alvin Burt), Letter to Austin Burt, Mar. 31, 1920, author's collection.

[60] Wm. A. Burt, Letter to Alvin Burt, May 23, 1844, Burt Papers.

[61] "Marquette Iron-Bearing District of Michigan," *U.S. Geological Survey (1897)*, p. 17.

[62] Wm. Johnston, Surv. Genl., *Annual Report for 1844*, to Thomas Blake, Comm. GLO, U.S. Cong., Senate (28-2), Oct. 30, 1844, S. Doc. 7, U.S. Serial 449, pp. 47–49.

[63] Wm. Johnston, Surv. Genl., Instructions to Douglass Houghton, July 10, 1844, Michigan Dept. of State, State Archives.

[64] John Bartlow Martin, *Call It North Country, The Story of Upper Michigan* (New York: Knopf, 1949), p. 48.

[65] Edsel K. Rintala, "The Geological Survey, 1841–1845," *Douglass Houghton: Michigan's Pioneer Geologist* (Detroit: Wayne Univ. Press, 1954), p. 80.

[66] Jacob Houghton, Letter to the editor of the *Mining Journal*, Oct. 16, 1869, cited in *Plain Dealer—Extra*, Nov. 20, 1869, Burt Papers. Letters from H. Mellen (Nov. 1, 1869) and Wm. Ives (Nov. 8, 1869), which describe the survey, are included. Some modern day authors have incorrectly placed Dr. Houghton with the survey party as late as Sept. 19, 1844, and Dr. Houghton even tried to convey the impression he was with Burt's party much longer than he actually was. (See Martin, p. 61). Houghton's brother Jacob, however, left no doubt when he wrote, "Dr. Douglass Houghton. . . confined his examinations mostly to the copper region, and until our return to Detroit in the fall of 1844, he was not aware of the existence of the iron deposits. This I had from his own lips."

[67] Wm. A. Burt, Letter to Phebe Burt, Aug. 14, 1844, Burt Papers.

[68] Field notes of William Austin Burt, Sept. 1844 survey, located in vol. 294, book 30028, Michigan Dept. of Natural Resources office, Lansing.

[69] Cited in Brooks, II, Appendix D, p. 236.

[70] Jacob Houghton, Letter to the editor of the *Mining Journal*, Oct. 16, 1869; Harlan Hatcher, "Discoveries on the Marquette Range," *A Century of Iron and Men* (New York: Bobbs-Merrill, 1950), pp. 21–37.

[71] Field notes of William Austin Burt, Sept. 1844 survey, pp. 30–32. The comment, "Spaltoric and Haemaltic Iron ore abound on this line," may have been added later by Dr. Houghton.

[72] Personal accounts of Burt's 1844 survey are printed in *Plain Dealer—Extra*, Nov. 20, 1869, Burt Papers.

[73] Ralph D. Williams, *The Honorable Peter White* (Cleveland: Penton, 1905), p. 19.

[74] Wm. A. Burt, Letter to Alvin Burt, Nov. 14, 1844, Burt Papers.

[75] Brooks, II, Appendix D, p. 235.

[76] Walter Havighurst, *Land of Promise* (New York: The Macmillan Co., 1946), p. 151.

[77] Isaac Lippincott, *Economic Development of the United States* (New York: D. Appleton, 1928), p. 194.

[78] James Truslow Adams (ed), *Dictionary of American History* (New York: Scribner's Sons, 1976), III, 157. The U.S.S. Michigan was used extensively during the Civil War as the only naval vessel on the Upper Great Lakes, and remained in service for 75 years. In 1909, it was renamed the Wolverine, after the new cruiser Michigan was commissioned.

[79] *U.S. Geological Survey (1897)*, p. 56.

XI. 1845 and 1846:

[80] From *What is Civilization?*

[81] P.M. Everett, "Recollections of the Early Explorations and Discovery of Iron Ore on Lake Superior," *MPHC*, XI (1888), 161–74 (Everett does not mention the Chippewa Chief Marji Gesick in this personal account.); Ralph D Williams, pp. 19–23; Walter Havighurst (ed), "Philo Everett: Finding the Iron Hills," *The Great Lakes Reader*, (New York: Macmillan, 1966), pp.95–99.

[82] Dunbar, *Michigan: A History of the Wolverine State*, p. 367

[83] See Robert Traver, *Laughing Whitefish* (New York: McGraw-Hill, 1965). This exciting historical novel is the story of efforts by Chief Marji-Gesick's daughter, Laughing Whitefish, to "collect on a debt owed her father" for his share in the Jackson iron mine. (For actual legal basis see "Compo v. Jackson Iron Co." (Supreme Court of Michigan), June 13, 1883, *The Northwestern Reporter*, XVI, 295–304, and "Kobogum et al v. Jackson Iron Co." (Supreme Court of Michigan), Oct. 25, 1889, *The Northwestern Reporter*, XXXXIII, 602–06).

Traver (John D. Voelker) describes William Austin Burt as "angular, beardless, dour-looking, with strong evangelical eyes, a member of that curious new American breed, the practical idealist, the self-educated jack-of-all-trades, the dreaming Yankee visionary who yet kept his eyes on the main chance." (p. 64).

[84] Willis Frederick Dunbar (ed), "Jackson Mine," *Michigan's Historical Markers* (Lansing: Michigan Historical Commission, 1967), p. 2.

[85] R[ay] A. Brotherton, "Discovery of Iron Ore: Negaunee Centennial (1844–1944)," *Michigan History* (1944), XXVIII, pp. 199–213.

[86] During the author's visit to Negaunee in 1977, Frank Matthews, colorful curator of the Jackson Mine Museum, graciously shared many Upper Peninsula stories with him, including the legend of the pine stump.

[87] Wm. A. Burt, Letter to Alvin Burt, November 14, 1844, Burt Papers.

[88] Wm. A. Burt, Letter to Samuel Williams, Chief Clerk, GLO, Jan. 30, 1845, Burt Papers.

[89] Lucius Lyon, Letter to John Mullett, April 18, 1845, "Letters of Lucius Lyon," *MPHC* (1896), XXVII, 599. Lyon was appointed Surveyor General of Michigan, June 23, 1845.

[90] Rintala, p. 81; White, p. 102.

[91] Wm. A. Burt, Letter (draft) to William J. Young , Apr. 9, 1845, Burt Papers.

[92] Wm. A. Burt, Letter (draft) to Alvin Burt, Apr. 9, 1845, Burt Papers.

[93] Lucius Lyon, Letter to John Mullett, April 18, 1845, "Letters of Lucius Lyon," *MPHC* (1896), XXVII, 599.

[94] Thayer, pp. 404–11.

[95] Wm. A. Burt, Letter (draft) to William J. Young , Apr. 9, 1845, Burt Papers.

[96] U.S. Cong., Senate, *Charles T. Jackson's Reports on The Linear Surveys*, 31-1, U.S. Serial 551, S. Doc. 1 (1849), p. 835.

[97] "Story of the Drowning of Dr. Douglass Houghton and Sketch of Peter McFarland, the Last Survivor of the Expedition," *MPHC*, XX, 662–66; U.S. Cong., House, Lucius Lyon, *Report of the Surveyor General of Ohio, Indiana, and Michigan*, to James Shields, Commissioner, G.L.O., Nov. 10, 1845, 29-1, U.S. Serial Set 481, Exec. Doc. 12, pp. 43–48.

[98] Dr. Houghton's geological notes from 7 townships were lost.

[99] Bernard C. Peters (ed), *Bela Hubbard's Account of the 1840 Houghton Exhibition* (No. Michigan Univ. Press, 1983), p. 5.

[100] J. Houghton, Jr. and T.W. Bristol, *Reports of Wm. A. Burt and Bela Hubbard, Esqs, on the Geography, Topography and Geology of the U.S. Surveys of the Mineral Region of the South Shore of Lake Superior, for 1845* (1846). This booklet also includes a list of 104 mining company leases made up to mid-1846.

[101] U.S. Cong., Senate, *Charles T. Jackson's Reports on The Linear Surveys*, 31-1, U.S. Serial 551, S. Doc. 1 (1849), p. 820.

[102] White, p. 110; U.S. Cong., House, Lucius Lyon, *Report of the Surveyor General of Ohio, Indiana, and Michigan*, to James Piper, Commissioner, G.L.O., Nov. 7, 1846, 29-2, House Doc. 9 (1846), p. 41.

[103] Iron ore production from the Menominee iron range began in 1877. Seven years later iron ore production began on Michigan's Gogebic range.

[104] U.S. Cong., Senate, *Charles T. Jackson's Reports on The Linear Surveys*, 31-1, U.S. Serial 551, S. Doc. 1 (1849), p. 852

[105] U.S. Cong., Senate, *Geological Report of Wm. A. Burt, on surveys of township lines, in 1846*, 31-1, U.S. Serial 551, S. Doc 1 (1849), pp. 842–75.

[106] Wm. A. Burt, Letter to Lucius Lyon, August 28, 1846, Burt Papers.

[107] *Lake Superior News* [Mackinaw, Mi.], July 11, 1846, cited in Everett N. Dick, *The Lure of the Land* (Univ. of Nebraska Press, 1970), pp. 29–30.

[108] "Pioneer Architecture in Southern Michigan", *Detroit News*, Nov. 6, 1931, Burt Papers, Burton Hist. Coll., Detroit Pub. Library.

[109] William Burt's house was destroyed by a tornado during the 1960's. The remaining "chalets" are located on 28 Mile Road in Romeo, Mi.

[110] Alvin Burt, Letter to Wm. A. Burt, Nov. 10, 1845, Burt Papers. George W. Jones (1804–1896) was reappointed to the office of Surveyor General of Iowa by President Polk in 1845. He had served in the U.S. Congress as a territorial representaive of Michigan (January 1835–April 1836) and Wisconsin (April 1836–January 1839). He was elected as a U.S. Senator from Iowa in 1848 and was re-elected in 1852.

[111] John Burt, Letter to Wm. A. Burt, Dec. 10, 1846, Burt Papers.

[112] U.S. Cong., Senate, *Geological Report of Bela Hubbard on Subdivisional Surveys by Dr. Houghton, in 1845,* 31-1, U.S. Serial 551, S. Doc. 1 (1849), p. 841.

[113] U.S. Cong., House, Lucius Lyon, *Report of the Surveyor General of Ohio, Indiana, and Michigan,* to James Shields, Commissioner, G.L.O., 29-1, , U.S. Serial Set 481, H. Doc. 12 p. 45.

[114] William Burt, Letter to Alvin Burt, Apr. 30, 1846, Burt Papers.

[115] Wm. J. Young, Letter to Lucius Lyon, Dec. 15, 1845, and April 6, 1846, Lyon Papers, Wm. L. Clements Library, Ann Arbor.

[116] Wm. J. Young, Letter to Lucius Lyon, April 6, 1846, Lyon Papers.

[117] Wm. J. Young, Letter to Lucius Lyon, May 2, 1846, Lyon Papers.

[118] Wm. J. Young, Letter to Lucius Lyon, Dec. 21, 1846, Lyon Papers.

[119] Wm. J. Young, Letter to Lucius Lyon, July 1, 1847, Lyon Papers.

[120] Wm. J. Young, Letter to Wm. A. Burt, June, 1844, Burt Papers.

[121] Wm. J. Young, Letter to Wm. A. Burt, Apr. 4, 1847, Burt Papers.

XII. 1847 and 1848:

[122] James Piper, Acting Commissioner, GLO, Special Instructions to Wm. A. Burt, May 1, 1847, Michigan State Archives, Lansing. Burt's contract is dated April 27, 1847.

[123] Approved by Congress Aug. 6, 1846.

[124] George H. Cannon, "Our Western Boundary," *MPHC* (1906), XXX, 244–61; George H. Cannon, "Michigan's Land Boundary," *MPHC* (1912), XXXVIII, 163–68.

[125] James H. Piper, Acting Commissioner, Letter to Lucius Lyon, Surv. Genl., Sept. 15, 1846, Michigan Dept. of State, State Archives.

[126] E. C. Martin, "Leaves From An Old Time Journal," *MPHC* (1906), XXX, 407.

[127] Wm. A. Burt's Field Notes on the Michigan-Wisconsin boundary survey, Michigan State Archives, Lansing.

[128] Wm. A. Burt's Field Notes, cited by Cannon, "Our Western Boundary," p. 258.

[129] *Mining Journal* [Marquette], Nov. 19, 1964. The remains of this tree were located, and the section marked by Burt was donated to the Marquette County Historical Society.

[130] Elizabeth S. Adams, "The State Line Historical Site," *Michigan History Magazine,* Dec., 1964, pp. 373–74.

[131] J. A. Williamson, Commissioner, GLO, Letter to John Burt, Feb. 20, 1878, cited in Hiram A. Burt, *Facts Upon Which is Founded the Claim of the Heirs of Wm. A. Burt, Deceased* (1886), p. 15.

[132] White, p. 110.

[133] U.S. Cong., Senate, *Field notes of George O. Barnes,* Sub-agent for J.D. Whitney, 31-1, U.S. Serial 551, S. Doc. 1 (1849), p. 629.

[134] Chas. Noble, Letter to J. Butterfield, Feb. 9, 1852, SGO Ltrs. Sent, National Archives.

[135] Cited in Stewart, p. 58.

[136] Cited in Stewart, p. 84.

[137] U.S. Cong., Senate, *Field notes of W. Gibbs,* Sub-agent for J.D. Whitney, 31-1, U.S. Serial Set 551, S. Doc. 1 (1849), p. 717.

[138] Cannon, "Our Western Boundary," p. 260.

[139] U.S. Cong., Senate, *Field notes of W. Gibbs,* Sub-agent for J.D. Whitney, 31-1, U.S. Serial Set 551, S. Doc. 1 (1849), p. 719.

[140] U.S. Cong., Senate, Field notes of W. Gibbs, Sub-agent for J.D. Whitney, 31-1, U.S. Serial Set 551, S. Doc. 1 (1849), p. 740. The remains of a hemlock bearing tree Burt marked in September, 1847, was located 130 years later near the present Gogebic County airport,north of Ironwood. The markings were clearly visible.

[141] U.S. Cong., Senate, Lucius Lyon, *Report of the Surveyor General of Ohio, Indiana, and Michigan,* to Richard Young, Comm., GLO, Nov. 30, 1847, 30-1, U.S. Serial Set 504, S. Doc. 2, p. 75.

[142] U.S. Cong., Senate, Lucius Lyon, *Report of the Surveyor General of Ohio, Indiana, and Michigan,* to Richard Young, Comm., GLO, Nov. 24, 1848, 30-2, U.S. Serial Set 539, S. Doc. 12, p. 31.

[143] U.S. Cong., Senate, Lucius Lyon, *Report of the Surveyor General of Ohio, Indiana, and Michigan,* to Richard Young, Comm., G.L.O., Nov. 30, 1847, 30-1, U.S. Serial Set 504, S. Doc. 2, p. 77.

[144] John Burt, "Autobiography," p. 3.

[145] Wm. J. Young, Letter to Wm. A. Burt, Mar. 5, 1848, Burt Papers.

[146] Wm. J. Young, Letter to Wm. A. Burt, Apr. 22, 1848, Burt Papers.

[147] Surveying and Mapping (1958), XVIII.

[148] Wm. J. Young, Letter to Lucius Lyon, Nov. 6, 1848, Lyon Papers.

[149] Wm. A. Burt, Petition to U.S. Congress, Jan. 4, 1850; cited in Hiram Burt, p. 5.

XIII. A Matter of Conscience:

[150] From *Table Talk*.

[151] Wm. E. Barton, The Life of Abraham Lincoln (Bobbs Merrill, 1925), I, 294.

[152] Abraham Lincoln, Letter to R. W. Thompson, May 25, 1849, cited in Archer Hayes Shaw (ed), The Lincoln Encyclopedia (New York: MacMillin, 1950), pp. 132–33; Also see Marian Clauson, *The Bureau of Land Management* (New York: Praeger, 1971), p. 29.

[153] U.S. Cong., Senate, Lucius Lyon, *Report of the Surveyor General of Ohio, Indiana, and Michigan*, to J. Butterfield, Comm., GLO, Nov. 5, 1849, 31-1, S. Doc. 1, U.S. Serial 550, pp. 282–85.

[154] Wm. A. Burt, Letter to John Wilson, Chief Clerk, GLO, Dec. 14, 1849, Burt Papers.

[155] Nicholson was a member of Wm. A. Burt's first survey crew in 1833. See Austin Burt, "Reminiscences," p. 38.

[156] U.S. Cong., Senate, Lucius Lyon, *Report of the Surveyor General of Ohio, Indiana, and Michigan*, to J. Butterfield, Comm., GLO, Nov. 5, 1849, 31-1, S. Doc. 1, U.S. Serial 550, pp. 308–09.

[157] Affidavit, signed by Burt's chainmen Geo. H. Cannon and Allen Church, sent by Lucius Lyon to Comm. Butterfield, Dec. 1849, SGO Ltrs. Sent, National Archives Record Group 49, roll 10, pp. 82–83.

[158] Surveyor John Brink originally surveyed the lake in 1839 and called it "Forbinson Lake." Wm. A. Burt's field notes of June 1849 appear to be the first official reference to the name "Higgins Lake". (Walter Romig, *Michigan Place Names*, p. 245.)

[159] Wm. A. Burt, Letter to Lucius Lyon, June 3, 1849, Lyon Papers.

[160] U.S. Cong., Senate, Lucius Lyon, *Report of the Surveyor General of Ohio, Indiana, and Michigan*, to J. Butterfield, Comm., GLO, Nov. 5, 1849, 31-1, S. Doc. 1, U.S. Serial Set 550, p. 284.

[161] Lucius Lyon, Letter to William Burt, Oct. 1, 1849, SGO Ltrs. Sent, National Archives Record Gp. 49, roll 10, p. 30.

[162] Lucius Lyon, Letter to John Norvell, U.S. District Atty., Nov. 10, 1849, SGO Ltrs. Sent, National Archives Record Group 49, roll 10, pp. 73–74.

[163] W. A. Burt, Field notes of "Explorations & Examinations", May 23 to Aug. 9, 1849, Michigan Archives.

[164] U.S. Cong., Senate, Lucius Lyon, *Report of the Surveyor General of Ohio, Indiana, and Michigan*, to J. Butterfield, Comm., GLO, Nov. 5, 1849, 31-1, S. Doc. 1, U.S. Serial Set 550, p.284.

[165] Lucius Lyon, Letter to John Norvell, U.S. District Atty., Nov. 10, 1849, SGO Ltrs. Sent, National Archives Record Group 49, roll 10, pp. 73–74.

[166] Harold H. Dunham, *Government Handout* (New York: Edwards Brothers, 1941), pp. 246–253.

[167] Lucius Lyon, Letter to John Norvell, U.S. District Atty., Nov. 7, 1849, SGO Ltrs. Sent, National Archives Record Group 49, roll 10, p. 69. The Act of Aug. 8, 1846, requires that suit be brought against any deputy surveyor whose work is found to be fraudulent, and "the institution of such suit shall act as a lien upon any property owned or held by such deputy, or his sureties, at the time such suit was instituted."

[168] George H. Cannon, "Our Western Boundary," *MPHC*, XXX (1906), 260.

[169] Wm. A. Burt, Letter to Lucius Lyon, Nov. 9, 1849, SGO Ltrs. Recd., National Archives Record Gp. 49.

[170] John Burt, *History of the Solar Compass* (Detroit: 1878), p. 9.

[171] The 1st session of the 31st Congress lasted 302 days, third longest of the 19th century.

[172] Wm. A. Burt, Letter to Lucius Lyon, Jan. 1, 1850, Lyon Papers, Wm. L. Clements Library, Univ. of Michigan, Ann Arbor.

[173] Wm. A. Burt, Letter (draft), addressee unknown, Burt Papers.

[174] These letter drafts are located in the Burt Papers.

[175] Everett Dick, *The Lure of the Land* (Lincoln: Univ. of Nebraska Press, 1970), p. 24.

[176] Wm. A. Burt, Letter (draft) to John Wilson, Clerk of Surveys, Dec. 20, 1849, Burt Papers.

[177] Wm. A. Burt, Letter (draft) to John Wilson, Clerk of Surveys, Dec. 20, 1849, Burt Papers.

[178] U.S. Cong., House, Charles Noble, Surv. Genl., *Report of the Surveyor Northwest of the Ohio*, Nov. 12, 1852, 32-2, U.S. Serial 673, H. Doc. 1, pp. 160–61.

[179] Thomas Simpson, "The Early Government Land Survey in Minnesota West of the Mississippi River," *Minnesota Historical Society Collections* (1899), X, 62.

[180] Justin Butterfield, Comm., GLO, Letter to C. E. Booth, Surv. Genl., Iowa, Dec. 18, 1849, GLO Ltrs. Sent, National Archives Records, M27, roll 13: pp. 301–02.

[181] Ex-Senator Alpheus Felch, *Solar Compass: Recent Statements in Support of the Claim of the Heirs of Wm. A. Burt, Deceased, for the Use Thereof in the Public Land Surveys,* Feb. 14, 1889, Burt Papers.

[182] *U.S. Constitution,* Article I, Section 8, Clause 8.

[183] Robert Calvert (ed), *The Encyclopedia of Patent Practice and Invention Management* (N.Y.: Reinhart, 1964).

[184] Ex-Senator George W. Jones, *Solar Compass: Recent Statements in Support of the Claim of the Heirs of Wm. A. Burt, Deceased, for the Use Thereof in the Public Land Surveys,* Feb. 23, 1889, Burt Papers.

[185] Wm. A. Burt, Letter to Phebe Burt, Dec. 18, 1849, Burt Papers.

[185] Letter draft, possibly sent to geologist Bela Hubbard, Dec. 31, 1849, Burt Papers.

[186] Wm. A. Burt, Letter to Lucius Lyon, Jan. 1, 1850, Lyon Papers.

[187] Holman Hamilton, *Zachary Taylor* (N.Y.: Bobbs-Merrill, 1951), p. 378.

[188] Cited in Hiram A. Burt, *Facts Upon Which is Founded the Claim of the Heirs of Wm. A. Burt, Deceased, Against the United States Government For the Use of the Solar Compass in the Survey of the Public Domain* (1886), pp. 4–5.

XIV. Justice For All:

[1] Lincoln's *First Inaugural Address,* March 4, 1861.

[2] U.S. Cong., House, *History of the United States House of Representatives,* 89-1, H. Doc. 250, p. 40.

[3] Alvin M. Tosepy, Jr., *The American Heritage History of The Congress of the United States* (N.Y.: American Heritage, 1975), p. 208.

[4] Robert Calvert (ed), *The Encyclopedia of Patent Practice and Invention Management* (N.Y.: Reinhart, 1964), pp. 397–98; Silvio A. Bedini, Thinkers and Tinkers, pp. 349–51.

[5] U.S. Cong., Senate, *Memorial of William C. Poole,* Feb. 11, 1839, 25-3, U.S. Serial 340, S. Doc. 204.

[6] U.S. Cong., House, *Memorial of Moses Smith,* Jan. 3, 1832, 22-1, U.S. Serial 217, H. Doc. 32; Bedini, pp. 343–44.

[7] U.S. Congress, House, *Report on the Memorial of Moses Smith,* Jan. 15, 1833, 22-2, U.S. Serial 236, H. Rep. 84.

[8] Wm. A. Burt, Letter (draft) to John Wilson, Clerk, GLO, Feb. 18, 1850, Burt Papers, Marquette County Historical Society, Marquette, Michigan.

[9] John Burt, *History of The Solar Compass* (1878), p. 10.

[10] U.S. Cong., House, *1850 Patent Office Report,* 31-2, H. Doc. 32, p. 6.

[11] Wm. J. Young, Letter to Wm. A. Burt, Dec. 14, 1847, Burt Papers.

[12] Wm. A. Burt, Letter (draft) to un-named recipient, Dec. 31, 1849, Burt Papers; Chas. Whittlesey, Letter to Senator A. Felch, Jan. 26, 1850, cited in Hiram Burt, *Facts,* p. 6.

[13] Wm. A. Burt, Letter (draft) to un-named recipient, Dec. 31, 1849, Burt Papers.

[14] Wm. A. Burt, Letter (draft) to John Wilson, Clerk, GLO, Feb. 18, 1850, Burt Papers.

[15] Wm. J. Young, Letter to Wm. A. Burt, Mar. 12, 1850, Burt Papers.

[16] Wm. J. Young, Letter to Wm. A. Burt, May 11, 1850, Burt Papers.

[17] Wm. J. Young, Letter to Wm. A. Burt, Mar. 31, 1850, Burt Papers.

[18] Wm. Young, Letter to Wm. A. Burt, Dec. 9, 1850, Burt Papers.

[19] Wm. A. Burt, Letter to Phebe Burt, May 25, 1851, reprinted in *Michigan History Magazine* (Spring 1929), XIII, 270.

[20] J. Aylesworth Bailey advertisement, Burt Papers.

[21] Wm. A. Burt, Letter to Warren Isham, March 12, 1850, *Michigan Farmer* (April 1850), VIII, 4.

[22] Wm. A. Burt, Letter to Lucius Lyon, June 20, 1849, Lyon Papers.

[23] U.S. Cong., House, *Report of the Surveyor General Northwest of the Ohio,* Lucius Lyon to GLO Comm. John Wilson, Nov. 12, 1852, U.S. Serial 673, H. Doc. 1, p. 160.

[24] *Michigan-Wisconsin TourBook* (AAA: 1984 edition), p. 52.

[25] Charles Noble, Surv. Genl, Letter to J. Butterfield, Comm. GLO, Sept. 11, 1851, SGO Ltrs. Recd., p. 281.

[26] U.S. Cong., House, Charles Noble, *Report of the Surveyor General of Ohio, Indiana, and Michigan,* to J. Butterfield, Comm., GLO, Nov. 12, 1850, 31-2, H. Doc. 9, U.S. Serial Set 597, pp. 67–77; Cannon, "Life and Times of William Austin Burt", p. 22.

[27] Leander Chapman, Surv. Genl., Letter to John Wilson, GLO Comm., Nov. 9, 1853, SGO Ltrs. Sent, National Archives.

[28] George H. White, "A Sketch of Lucius Lyon, One of the First Senators from Michigan," *MPHC* (1888), XIII, 332–33.

[29] Chas. Noble, Surv. Genl., Letter to J. Butterfield, GLO Comm., Mar. 13, 1850, GLO Ltrs. Sent, National Archives.

[30] Chas. Noble, Surv. Genl., Letter to Wm. A. Burt, July 13, 1850, SGO Ltrs. Sent., National Archives.

[31] Chas. Noble, Surv. Genl., Letter to J. Butterfield, Nov. 12, 1850, SGO Ltrs. Sent., National Archives.

[32] Chas. Noble, Letter to J. Butterfield, Dec. 14, 1850, SGO Ltrs. Sent, National Archives.

[33] Leander Chapman, Surv. Genl., Letter to John Wilson, GLO Comm., Jan. 18, 1854, Report of John L. Mitchell, SGO Ltrs. Sent, National Archives.

[34] Chas. Noble, Letter to John Wilson, GLO Comm., Oct. 14, 1851, GLO Ltrs. Sent, National Archives.

[35] Chas. Noble, Letter to John Wilson, GLO Comm., Nov. 17, 1852, GLO Ltrs. Sent, National Archives.

[36] John H. Forester, "Memoir of John Mullett," *MPHC*, VIII, 175–78.

[37] Michael D. Moore, *Michigan's Famous and Historic Trees* (Michigan Forest Assn., Feb., 1977), p.11.

[38] *The Mining Journal* [Marquette], Mar. 3, 1894, "Sketch of Austin Burt's Life". Austin Burt is credited with devising an improved method of subdividing townships in 1848, although his name does not appear on the proposal. It is reported that the government officers initially opposed the plan; it was, however, later accepted and practiced in all the new states and territories.

[39] Wm. A. Burt, Letter to Lucius Lyon, Surv. Genl., Jan. 7, 1849, Ltrs. Recd. by the Surv. Genl. NW of the Ohio, National Archives.

[40] John Burt, History of the Solar Compass, p. 5.

[41] U.S. Cong., House, *Report of Caleb H. Booth, Surveyor General for Iowa,* Nov. 11, 1850, 31-2, U.S. Serial 597, H. Doc. 9, p. 46.

[42] Wm. A. Burt, *A Key to the Solar Compass and Surveyor's Companion* (Philadelphia: Young, 1855), p. 54.

[42] Wm. Dewey and R. Walker, Letter to Messers. Hendershott and Minor, Sept. 30, 1850, cited in J.S. Dodds (ed), Original Instructions Governing Public Land Surveys 1815–1855 (Iowa: Dodds, 1944), pp. 460–63.

[43] Charles Whittlesey, Letter to Senator A. Felch, Jan. 26, 1850, cited in Hiram Burt, "Facts," pp. 6–7.

[44] U.S. Cong., Senate, *Foster & Whitney's Report of the Lake Superior Land District,* March 1851, 32-1, U.S. Serial 609, S. Doc 4.

[45] John Locke, Letter to John Wilson, Chief Clerk, GLO, April 18, 1850, cited in Hiram Burt, p. 7.

[46] Justin Butterfield, Comm., GLO, Letter to Senator A. Felch, April 2, 1850, cited in Hiram Burt pp. 5–6. Deseret was the name given to the large state proposed by Mormon leader Brigham Young. In 1850, however, Congress established the smaller state of Utah.

[47] U.S. Cong., House, J. Butterfield, Comm., GLO, *Annual Report for 1850,* to A.H. Stuart, Sect. of the Interior, Nov. 30, 1850, 31-2, U.S. Serial 597, H. Doc 9, p. 3.

[48] Wm. J. Young, Letter to Wm. A. Burt, Jan. 27, 185[1]; White, p. 114.

[49] Wm. J. Young, Letter to Wm. A. Burt, Dec. 9, 1850, Burt Papers.

XV. New Frontiers:

[50] From Sam Walter Foss, *The Coming American.*

[51] Cited in Robert F. Dalzall, Jr., *American Participation in the Great Exhibition of 1851* (Mass.:Amherst College Press, 1960), p. 28.

[52] Letter of introduction from Senator Cass, dated April 1, 1851., cited in "Letters Relative to William A. Burt, 1851–1854," *Michigan History Magazine,* Spring, 1929, XIII, 265–77.

[53] Horace Greely, Letter to W. H. Verity, 1854. Three years earlier John L. B. Soule, in a *Terre Haute Express* (Ind.) editorial, wrote, "Go west young man, go west."

[54] "Letters," *Michigan History Magazine,* Spring, 1929, XIII, 265–77.

[55] George H. Cannon, "The Life and Times of William A. Burt of Mt. Vernon, Michigan,", *MPHC*, V, 115–23.

[56] Anthony Hyman, *Charles Babbage* (Princeton Univ. Press, 1982), p. 221. The structural units of the Crystal Palace were prefabricated and shipped to the site for assembly. After the exhibition closed, the Crystal Palace was dismantled and re-erected at Sydenham. It remained a popular tourist attraction until 1936, when it was destroyed by fire.

[57] Elizabeth Longford, *Victoria, R.I.* (New York: Harper, 1964), p. 131.

[58] Wm. A. Burt, Letter to Phebe Burt, "Letters Relative to William A. Burt, 1851–1854," *Michigan History Magazine,* Spring, 1929, XIII, 265–77.

[59] Wm. A. Burt, Letter to Phebe Burt, May 25, 1851, "Letters," *Michigan History Magazine,* Spring, 1929, XIII, p. 270.

[60] Wm. A. Burt, Letter to Phebe Burt, May 25, 1851, "Letters," *Michigan History Magazine*, Spring, 1929, XIII, p. 271.

[61] Longford, p. 131.

[62] Cited in John Burt, *History of the Solar Compass* (1878), p. 3.

[63] Wm. A. Burt, Letter to Phebe Burt, June, 28, 1851, "Letters," *Michigan History Magazine*, Spring, 1929, XIII, pp. 272–73.

[64] Wm. A. Burt, Letter to Phebe Burt, July, 20, 1851, "Letters," *Michigan History Magazine*, Spring, 1929, XIII, p. 272–73.

[65] Three weeks after William Austin left England, the American yacht *America* beat Britain's *Titania* at the Royal Yacht Club regatta at Cowes to win the now famous "America's Cup" for the first time. The United States retained the coveted trophy for the next 132 years, when, in 1983, an Australian yacht defeated the American entry.

[66] Longford, p. 131.

[67] Glyndon G. Van Deusen, Horace Greeley: Nineteenth Century Crusader (Hill & Wang, 1964), p. 6.

[68] Cited in Farmer, p. 180.

[69] Wm. A. Burt, Letter to Phebe Burt, July, 20, 1851, "Letters," *Michigan History Magazine*, Spring, 1929, XIII, p. 273.

[70] Lola Cazier, *Surveys and Surveyors of the Public Domain, 1765—1975* (Washington, D.C.: GPO, 1976), p. 112.

[71] White, p. 114.

[72] Chas. Noble, Surv. Gen, Letter to J. Butterfield, Comm., GLO, Apr. 7, 1851, SGO Ltrs. Sent, National Archives.

[73] Justin Butterfield, Comm., GLO, Letter to John Preston, Surv. Genl., March 6, 1851, GLO Ltrs. Sent, National Archives.

[74] Wm. A. Burt, Letter to Senator A. Felch, Feb. 12, 1851, Felch Papers, Clements Library, Univ. of Michigan.

[75] Kimball Webster, "With Compass & Chain in Oregon," *The Gold Seekers of '49* (New Hampshire: Standard, 1917), p. 205.

[76] U.S. Cong., House, *Report of the Surveyor General of Oregon*, Oct. 20, 1851, 32–1, U.S. Serial 636, H. Doc 2, p. 193.

[77] U.S. Cong., Senate, Lieut. Henry L. Abbot, Corp. of Topographical Engineers, *Report Upon Explorations For a Railroad Route, from the Sacramento Valley to the Columbia River*, 1855, 33–2, S. Exec. Doc. 78, v. 6, Serial Set 763, p. 12.

[78] Chas. Noble, Surv. Gen, Letter to J. Butterfield, Comm., GLO, Apr. 7, 1851, SGO Ltrs. Sent, National Archives.

[79] Samuel King, Letter to J. Butterfield, Apr. 22, 1851, GLO Ltrs. Recd., National Archives.

[80] For a description of Richard Patten (1792–1865), see Charles E. Smart, The Makers of Surveying Instruments in America Since 1700 (New York: Regal Art Press, 1962), pp. 14–15. William J. Young was selling Burt Solar Compasses in 1852 for $200 to compete with Patten's price. He did not feel that Patten's workmanship matched his own, however. (Wm. J. Young, Letter to Wm. A. Burt, Apr. 10, 1852, Burt Papers). Young may have been correct, as few of Patten's instruments seem to have survived. (See Roger C. Gerry, "Richard Patten: mathematical instrument maker," *Antiques*, July 1959, pp. 56–58.)

[81] White, p. 114.

XVI. John Burt in the Upper Peninsula:

[82] William Wordsworth, *Poems Dedicated to National Independence*.

[83] Silas Farmer, *The History of Detroit and Michigan* (1889), p. 1185.

[84] On October 8, 1839, John's brother Austin married Lydia Calkins, Julia's sister.

[85] Burton H. Boyum (ed), *Mather Mine* (Marquette County Hist. Soc., 1979), p. 17; *Carp River Forge: A Report* (Michigan History Div., Michigan Dept. of State), pp. 8–9.

[86] Alan D. Strocke, "The Influence of John Burt on Marquette and the Upper Peninsula," Diss. Northern Michigan Univ., Burt Papers.

[87] Ralph D. Williams, *The Honorable Peter White, A Biographical Sketch of the Lake Superior Iron Country* (Cleveland: Penton, 1905), p. 46.

[88] Farmer, p. 1183. The members of the group included Zachariah Chandler, Henry N. Walker, Eber Brock Ward, and H.P. Baldwin.

[89] John Burt, "Material for Hon. John Burt's Autobiography," Burt Papers, Marquette County Historical Society, Marquette, MI, p. 6.

[90] "Early Iron Enterprises in Pennsylvania," *Tenth Census of the United States—Manufacturers,* June 1, 1880, p. 83.

[91] John Burt, "Autobiography," p. 6.

[92] U.S. Cong., Senate, *Foster and Whitney Mineral Reports,* to GLO Comm. J. Butterfield, Sept. 25, 1850, 31-2, U.S. Serial 597, S. Doc 9, p. 153. During 1850 five tons of ore were taken from the Jackson mine to Newcastle, Pennsylvania and made into blooms, or iron bars, for reworking.

[93] Strocke, p. 10.

[94] John Burt, Letter to William Burt, July, 1851, Burt Papers, cited in Strocke, pp. 9–10.

[95] John Burt, "Autobiography," p. 6.

[96] John N. Dickinson, "The Canal at Sault Ste. Marie, Michigan: Inception, Construction, Early Operation, and the Canal Grant Lands," Diss. Univ. of Wisconsin, 1968, p. 43.

[97] U.S. Cong., House, Chas. Noble, Surv. Genl., *Annual Report of the Surveyor General,* to J. Butterfield, Comm., GLO, Oct. 21, 1851, 32-1, U.S. Serial 636, H. Doc. 2, p. 50.

[98] Wm. Chandler, *Illustrated History of the St. Mary's Falls Ship Canal* (Chapman & Kibby, 1893), n. pag.

[99] John N. Dickinson, *To Build a Canal; Sault Ste. Marie, 1853–1854 and After* (Ohio State Univ. Press, 1981), p. 26.

[100] Irene D. Neu, "The Mineral Lands of the St. Mary's Falls Ship Canal Company," *The Frontier in American Development,* David M. Ellis (ed) (New York: Cornell Univ. Press, 1969), pp. 164–65.

[101] Ralph D. Williams, *The Honorable Peter White* (1905), p. 107.

[102] Cited in John Bartlow Martin, *Call It North Country* (New York: Knopf, 1949), p.113.

[103] William Chandler, *Illustrated History of the St. Mary's Falls Ship Canal* (Chapman & Kibby, 1893), n. pag.

[104] William Chandler, *History of the St. Mary's Falls Ship Canal* (1879), n. pag.

XVII. 1852—1855:

[105] John Wilson, Comm., GLO, Letter to David T. Disney, Chairman, Committee on Public Lands, Feb. 16, 1854, cited in Hiram Burt, pp. 13–14.

[106] "Survey of the Iowa-Minnesota Boundary Line," *Annals of Iowa,* XVI (7), January 1929, 3rd Series, 482–502; also see Surveyors General Letter Books, Sect. of State, Des. Moines, Iowa.

[107] Willis F. Dunbar, Michigan: A History of the Wolverine State (Grand Rapids: Eerdmans, 1966), pp. 308–09.

[107] John Wilson, Letter to George B. Sargent, Surv. Genl., GLO Ltrs. Sent, National Archives.

[108] "Survey of the Iowa-Minnesota Boundary Line," *Annals of Iowa,* XVI (7), January 1929, 3rd Series, 491.

[109] Roscoe L. Lokken, *Iowa Public Land Disposal* (State Hist. Soc. of Iowa, 1942), p. 62.

[110] George B. Sargent, Surv. Genl. for Wisc. & Iowa, Letter to J. Butterfield, July 17, 1852, Surveyors General Letter Books, Sect. of State, Des. Moines, Iowa.

[111] U.S. Cong., House, *Walbridge Report on the Solar Compass,* 34-1, H. Report 20, p. 3.

[112] Nathan Butler, Boundaries and Public Land Surveys of Minnesota," *Minnesota Historical Society Collections* (1908) XII, 650.

[113] J. W. Foster, Letter to J. Butterfield, Feb. 17, 1852, cited in Hiram Burt, *Facts Upon Which is Founded the Claim of the Heirs of Wm. A. Burt, Deceased, Against the United States Government* (1886), p. 7.

[114] J. Butterfield, Letter to Senate Comm. on Public Lands, Mar. 3, 1852, GLO Ltrs. Sent, National Archives; also cited in John Burt, *History of the Solar Compass* (Detroit: Gulley's Presses, 1879), pp. 7–8.

[115] J. Butterfield, Letter to Senate Comm. on Public Lands, Mar. 3, 1852, GLO Ltrs. Sent, National Archives.

[116] John Wilson, Chief Clerk, GLO, Letter to Senator A. Felch, Mar. 4, 1852, cited in Hiram Burt, *Facts,* pp. 9–10.

[117] William Austin had just returned from assisting his son Austin on a correctional survey near Midland in the Lower Peninsula. At the same time Wells Burt was engaged in correctional work near Grand Marais in the Upper Peninsula. See Charles Noble, Letter to Wm. A. Burt, April 15, 1852, SGO Ltrs. Sent, National Archives.

[118] George H. Cannon, Letter to Peter White, May 15, 1905, cited in Charles Moore (ed), *The Saint Mary's Falls Canal* (Detroit, 1907), p. 82.

[119] George H. Cannon, "The Life and Times of William A. Burt of Mt. Vernon, Michigan," *MPHC,* V, 115–23.

[120] Strocke, p. 30.

[121] Dickinson, p. 47.

[122] Williams, p. 110.

[123] *Michigan House Journal,* 1853, p. 37.

[124] Wm. A. Burt, Letter to John Burt, February 14, 1853, (from Lansing), Burt Papers.

[125] John Burt, Letter to Wm. A. Burt, Jan. 3, 1853, Burt Papers; Strocke, p. 31.

[126] John Burt, Letter to Wm. A. Burt, Jan. 3, 1853, Burt Papers.

[127] John Burt, Letter to his brother (probably Wells or Austin), Feb. 7, 1853, Burt Papers.

[128] Wm. J. Young, Letter to John Burt, Feb. 11, 1853, Burt Papers.

[129] John Burt, Letter to Wm. J. Young, Feb. 17, 1853, Burt Papers.

[130] U.S. Cong., Senate, "Borland Report," *Senate Journal,* 32-2, Rep. Com. 429.

[131] Wm. A. Burt, Letter to John Burt, Feb. 14, 1853, Burt Papers.

[132] S.P. Ely, "Historical Address," July 4, 1870, *MPHC,* VII, 168; John Burt obituary, August 16, 1886, Burt Papers.

[133] *A Trip Through the Lakes of North America Embracing a Full Description of the St. Lawrence River* (New York: J. Disturnell, 1857), pp. 66–68.

[134] Peter White, "The Iron Region of Lake Superior," *MPHC* (1885), VIII, 155.

[135] John Burt, Letter to Wm. Burt, May 30, 1852, Burt Papers, cited in Strocke, p. 11.

[136] John Burt, "Autobiography," p. 10; T. B. Brooks, Geological Survey of Michigan: Upper Peninsula, 1869–1873 (New York: Bien, 1873), I, 22.

[137] Brooks, I, 22.

[138] Tower Jackson, Letter to officials of the Cleveland Iron Mining Co., Dec 1853, cited by Harlan Hatcher, *A Century of Iron and Men* (New York: Bobbs-Merrill, 1950), p. 64.

[139] John Burt, "Autobiography," p.10.

[140] Margaret Pilkington, Letter to Wm. A. Burt, Mar. 22, 1854, cited in Michigan History Magazine (Spring, 1929), XIII, 274–76.

Note by Horace E. Burt, *Michigan History Magazine* (Spring, 1929), XIII, 276–77.

[141] Brooks, I, 30.

[142] *The Mining Journal* (Marquette), March 3, 1894.

[143] Kenneth D. LaFayette, Flaming Brands: Fifty Years of Iron Making in the Upper Peninsula of Michigan, 1848–1898 (Marquette: Northern Michigan Univ., 1977), pp.19–20.

[144] Charles K. Hyde, *The Upper Peninsula of Michigan: An Inventory of Historic Engineering and Industrial Sites* (Wash., D.C.: U.S. Dept. of the Interior, GPO, 1978), pp. 70–73.

[144] John Burt, "Autobiography," p. 13.

[145] Peter White, "The Iron Region of Lake Superior," MPHC, p. 160.

[146] *The Congressional Globe,* March 9, 1854. Senate debates, pp. 581–82.

[147] Thomas Simpson, "The Early Government Land Survey in Minnesota West of the Mississippi River," *Minnesota Historical Society,* X, 66.

[148] *The Congressional Globe,* March 9, 1854. Senate debates, pp. 581–82.

XVIII. 1855:

[1] From *Works,* Book II, Ch. 48.

[2] S.P. Ely, "Historical Address," July 4, 1870, Michigan Pioneer Historical Collections (40 volumes, Lansing: 1877–1929), VII, 168; hereafter cited as *MPHC.*

[3] Peter White, "The Iron Region of Lake Superior," *MPHC* (1885), VIII, 154.

[4] John N. Dickinson, *To Build a Canal; Sault Ste. Marie, 1853–1854 and After* (Ohio State Univ. Press, 1981), p. 106. Brooks later served as president of the Michigan Central Railroad from 1856 to 1867.

[5] Cited in Alan D. Strocke, "The Influence of John Burt on Marquette and the Upper Peninsula," Diss. Northern Michigan Univ., p. 35, Burt Papers, Marquette County Historical Society, Marquette, Michigan.

[6] James R. Barry, *Ships of the Great Lakes—300 Years of Navigation* (Howell-North, 1973), p. 76.

[7] John Burt's obituary, *Detroit Free Press,* Aug. 1886, Burt Papers.

[8] Ralph D. Williams, *The Honorable Peter White* (Cleveland: Penton, 1905), p. 116; A.O. Backert, The ABC of Iron & Steel (Penton, 1921) 4th ed., p.38.

[9] Dr. Shoenberger did not live to see the completion of the Soo Canal. He died in 1854, at age 73.

[10] Peter White, p. 152.

[11] *Lake Superior Journal* Advertisement, Oct. 5, 1854, cited in Strocke, pp. 15–16.

[12] John Burt, "Autobiography," p. 12.

[13] *Mining Journal*, December 1, 1885, cited in Strocke, p. 16.

[14] Leonard N. Neitz, Letter to Wm. A. Burt, Mar. 6, 1854, Burt Papers.

[15] Wm. A. Burt, Letter to John Wilson, Comm., GLO, Sept. 20, 1852, Burt Papers.

[16] Chas. Noble, Surv. Genl., Letter to John Wilson, Aug. 24, 1852, SGO Ltrs. Sent, National Archives.

[17] Agreement of Dissolution of Burt & Bailey, Feb. 13, 1856, Burt Papers.

[18] The Establishments of Burt & Bailey (1853–1856), J.& W. Burt (1856–1857), and Grant & Crosman (1858–1861) were located at 214 Jefferson Avenue, Detroit, MI. Within two months after the death of William A. Burt, his son William sold his interest in manufacturing solar compasses, and no member of the Burt family engaged in that end of the business again.

[19] Charles E. Smart, *The Makers of Surveying Instruments in America Since 1700* (N.Y.: Regal Art Press, 1962), p. 25.

[20] Smart, p. 53.

[21] Act of Congress, approved May 30, 1862 (Sec. 2399. R.S.)

[22] According to John Burt the solar compass was "adopted as a standard instrument for public surveys by government officials in 1852." ("Autobiography", p. 15, Burt Papers) It has been inaccurately printed that "The solar compass did not come into general use for nearly 50 years." (BLM brochure, Restoration of Lost or Obliterated Corners, first printed in the 1939 revision and repeated in subsequent editions.)

While use of the magnetic compass was not entirely prohibited until 1894, the solar compass was in general use on the principal lines of the public land surveys during the 1850's. The 1868 W.& L.E. Gurley Manual of Surveying Instruments, in describing the solar compass, notes "It has since come into general use in the surveys of U.S. Public Lands, the principal lines of which are required to be run with reference to the true meridian." Thomas Donaldson, in Public Domain, by Act of Congress, 1880, wrote, "The instruments employed in the full-work by United States surveyors consist of solar compasses, transits, and common compasses of approved construction."

In correspondence to the author, July 8, 1979, Minnesota Land Surveyor Carlisle Madson wrote, "When a method, procedure or call for equipment first appeared it immediately came into general use. It continued in general use until it was deleted from the manual of instructions. Obsolete or superseded instructions were not repeated in subsequent manuals. . . . Since the 1855 Manual of Instructions was made a part of the surveying contracts—by lawmthere is no reason to question the general usage of Burt's solar compass in surveying principal lines and township lines, or in the survey of other lines (subdividing and meandering) where the ordinary compass was not reliable."

[23] W. D. Jones, "Comments on the General Instructions", J. S. Dodds (ed), *Original Instructions Governing Public Land Surveys 1815–1855* (Iowa: Dodds, 1944), p. 476.

[24] Dodds, p. 1.

[25] Cited in Lola Cazier, *Surveys and Surveyors of the Public Domain: 1785–1975* (Wash.: GPO, n.d.), p. 64.

[26] Nathan Butler, "Boundaries and Public Land Surveys of Minnesota," *Minnesota Historical Society Collections*, vol. XII (Dec.,1908), p. 659–60.

[27] Robert P. Multhauf, "Early Instruments in the History of Surveying: Their Use and Their Invention," *Surveying and Mapping* (1958), XVIII, 412.

[28] Wm. A. Burt, Letter to an unspecified addressee, May 10, 1855, Burt Papers.

XIX. 1856:

[29] The publisher William C. Young and the instrument-maker William J. Young were probably related to each other.

[30] *Walden,* XVIII, Conclusion.

[31] Wm. A. Burt, Letter to Stansbury & Pitman, patent agents, Jan. 2, 1858, Burt Papers.

[32] Wm. J. Young, Letter to Wm. A. Burt, Apr. 10, 1855, Burt Papers.

[33] Wm. J. Young, Letter to Wm. A. Burt, Aug. 20, 1855, Burt Papers.

[34] Wm. A. Burt's handwritten notes, Michigan Society of Registered Land Surveyors historical collection.

[35] Matthew Fontaine Maury (1806–1873) was in charge of the Navy's Depot of Charts and Instruments (later the U.S. Naval Observatory and Hydrographical Office) from 1842 to 1861. His wind and current charts that are still used today were first prepared in 1847. By 1855 the use of Maury's charts had helped to shorten the travel time of vessels traveling from New York to San Francisco by 25%. In 1855 Maury published the classic text on oceanography, "The Physical Geography of the Sea."

[36] M. F. Maury, Letter to Commodore Jos. Smith, Acting Chief of Ordinance and Hydrography, March 11, 1856, cited in Wm. A. Burt, *Description and Use of the Equatorial Sextant* (1858), p. 10.

[37] Maury's increased fame during the 1850's widened the intense rivalry that existed between Maury and A. D. Bache, Supt. of the Coast Survey. Bache feared that control of the Coast Survey would be transferred to Maury and the Navy Department. (See Francis Leigh Williams, *Matthew Fontaine Maury, Scientist of the Sea* (New Jersey: Rutgers Univ. Press, 1963), p. 235.)

[38] A. D. Bache, Letter to John Burt, March 29, 1856, cited in Wm. A. Burt, *Description and Use of the Equatorial Sextant* (1858), p. 11.

[39] Wm. A. Burt, Letter to Phebe Burt, Aug. 15, 1856, and Letter to Prof. Chas. Page, Aug. 19, 1856, Burt Papers.

[40] Wm. A. Burt, Letter to Alfred Young, son of Wm. J. Young, Nov. 2, 1856, Burt Papers.

[41] Blueprints of foreign patents, Burt Papers.

[42] U.S. Cong., House, *Solar Compass,* March 31, 1856, 33-1, H. Report 20.

[43] U.S. Cong., House, *Reports and Debates in the Thirty-third and Thirty-fourth Congresses on the Solar Compass,* July 26, 1856, p. 29.

[44] Stewart, pp. 91-92.

[45] William B. Edwards, *The Story of Colt's Revolver* (Pennsylvania: Stackpole, 1953), Ch. 45: "Patent Extension"; U.S. Cong., House, Samuel Colt, Extension of Patent, H. Rpt. 132, 33–2, p. 808. Colt's effort was unsuccessful.

XX. 1857 and 1858:

[46] David Morton, *Who Walks with Beauty*

[47] Strocke, p. 37.

[48] John Burt, *1856 Report of the Superintendent of the St. Mary's Falls Ship Canal*. By 1860, the tonnage of iron ore shipments through the Soo Canal gates increased tenfold.

[49] Dickinson, p. 131.

[50] "History of the St. Mary's Falls Ship Canal," 1878 Annual Report of the Superintendent and Collector of the St. Mary's Falls Ship Canal, p. 8.

[50] Biographical sketch of Elisha Calkins, *History of Macomb County, Michigan* (1882), pp. 649–50.

[51] Elisha Calkins, *Report of St. Mary's Falls Ship Canal,* to Gov. Kinsley S. Bingham, Dec. 28, 1857. This previously unpublished report was printed in Philip M. Mason, "The Operation of the Sault Canal, 1857," *MPHC* (1955), XXXIX, 71–79.

[52] John Burt, "Report of the Superintendent of the St. Mary's Falls Ship Canal", to Gov. Bingham, Dec. 20, 1857.

[53] Brooks, p. 22; S. P. Ely, p. 168.

[54] John Burt, "Autobiography," p. 11.

[55] John Burt, "History of the Solar Compass," pp. 11–12.

[56] William A. Burt, patentee, *Description and Use of the Equatorial Sextant* (Detroit, Jan. 1, 1858).

[57] Wm. A. Burt, Letter to Charles Stansbury, October 3, 1857, Burt Papers.

[58] William A. Burt obituary, *Detroit Free Press,* August 20, 1858.

[59] Alfred Young, Letter to Wm. A. Burt, April 5, 1858, Burt Papers.

[60] Wm. J. Young, Letter to Wm. A. Burt, May 30, 1858, Burt Papers.

[61] Horace E. Burt, "Historical Notes," *Michigan History Magazine,* Spring 1929, XIII, pp. 368–70.

[62] Phebe Burt died in Detroit on August 23, 1864. She and William Austin are among 48 members of the Burt family buried at the Burt monument site in Detroit's historic Elmwood cemetery.

XXI. The Burts in the U.P.:

[63] Michel De Montaigne, *Essays,* Book I, Ch. xx.

[64] Farmer, p. 1182. Wells moved to Ypsilanti in 1865.

[65] Strocke, p. 18.

[66] The John Burt House is located at 220 Craig Street, Marquette, MI, 49855. (906) 226-6821. Open July–August, 7 days/wk., 9 a.m.–5 p.m. A plaque acknowledges the John Burt House as the first permanent building in Marquette, developed to further construction of the first stage route, first railroad, and the first charcoal furnace on the Marquette range. John Burt is recognized as the pioneer designer of the first iron ore pocket dock ever built. (See Kenyon Boyer, *Historical Highlights # 91* (Early Ore Docks), July 15, 1956.

[67] Strocke, p. 17.

[68] Ernest H. Rankin, "History of Burt, Mather Blocks In Marquette Told," *Mining Journal* (Marquette), August 12, 1964.

[69] Ernest Rankin, "Fire Claims John Burt's Generosity," *Inland Seas* (1965), pp. 84–85.

[70] Strocke, p. 20.

[71] S. P. Ely, "Historical Address [to the Pioneer Society of Michigan]," *MPHC*, VII, 178.

[72] "Pig iron" refers to the mass of iron that results from the reduction of iron ore in the blast furnace. It is cast into "pigs" that are used for making steel, cast iron, or wrought iron. The first pig iron in the Lake Superior region was produced in 1858 in a furnace on the Dead River northeast of Marquette.

[73] Kenneth D. LaFayette, *Flaming Brands: Fifty Years of Iron Making in the Upper Peninsula of Michigan*, 1848–1898, pp. 11–12. LaFayette provides the following definitions: Muck bar is "bar rolled from a squeezed bloom." A bloom is "an intermediate product which has been rolled or forged down from an ingot and is destined for further working into bars, sheet, tubes, and forgings, etc." (p. 48).

[74] Strocke, p. 22.

[75] *Mining Journal* (Marquette), Dec. 1, 1885; cited in Strocke, p. 22.

[76] *History of the Upper Peninsula*, p. 428.

[77] Brooks, p. 59.

[78] LaFayette, p. 29.

[79] Iron ore shipments from the Lake Superior Iron Company's mines during the period 1858–1900 totaled 8,850,309 gross tons. Shipments of iron ore from the Cleveland-Cliffs group of mines totaled 8,679,021 gross tons during the period 1854–1900. In 1897 the Cleveland-Cliffs Iron Company purchased a one-quarter interest in the Lake Superior Iron Company, while the remaining interest was purchased by the Oliver Iron Mining Company, later a division of the U.S. Steel Corporation. W. Clayton Burt, son of Wells Burt, was the only member of the Burt family still on the Board of Directors of the Lake Superior Iron Company in 1898. The company's properties continued to operate until 1946. See Van Hise, Leith, "The Geology of the Lake Superior Region," U.S. Geological Survey (1911), LII, 52–59; David T. Day, Mineral Resources of the United States—1901 (Wash.: GPO: 1902), pp. 62–64; J. E. Jopling, "A Brief History of the Cleveland-Cliffs Iron Company," *Michigan History Magazine*, V (1921), p. 162; J.E. Jopling, "Personal Experiences of a Mining Engineer," *Michigan History Magazine*, XI (1927), pp. 191–207; John Burt, "Autobiography," p. 12.

[80] LaFayette, p. 18; Burton H. Boyum, *The Saga of Iron Mining in Michigan's Upper Peninsula* (Marquette: Longyear Research Library, 1877), p.11.

[81] John Burt, "Autobiography," p. 14.

[82] Strocke, p. 23.

[83] *Michigan Official Directory and Legislative Manual* (1921 & 1922), p. 226. Charles M. Croswell was also chosen a Presidential elector at large.

[84] John Burt, "Autobiography," p. 15.

[85] Strocke, p. 25.

[86] Ernest H. Rankin, "Burt's Solar Compass," *Inland Seas* (n.d.), pp. 202–04.

[87] "Lawyers in the Lock-Pit," *Mining Journal* (Marquette), June 18, 1891, Burt Papers.

[88] Otto Fowle, *Sault Ste. Marie and It's Great Waterway* (New York: Putnam's Sons, 1925), p. 438.

[89] *Mining Journal* (Marquette), June 18, 1891, Burt Papers.

[90] *Mining Journal*, June 18, 1891.

[91] See "John Burt v. The United States," Case No. 13782, Cases decided in The Court of Claims at the Term of 1883–84, XIX, GPO, 1884, 120–23; "Lawyers in the Lock-Pit," *Mining Journal* (Marquette), June 18, 1894.

[92] Strocke, pp. 45–46.

XXII. The Final Chapter:

[93] *Familiar Studies of Men and Books*

[94] John Burt, *History of the Solar Compass* (Detroit: Gulley's Presses, 1878), p. 13.

[95] Hiram A. Burt, *Facts Upon Which is Founded the Claim of the Heirs of Wm. A. Burt, Deceased, Against the United States Government For the Use of the Solar Compass in the Survey of the Public Domain, and documents in support thereof* (Washington: Gray & Clarkson, 1886), 34 pages.

[96] John H. Mullett, "Invention of the Solar Compass," *Proceedings of the Michigan Engineering Society* (1888), 99–101.

[97] Cited in Horace E. Burt, "The Solar Compass-By Whom Invented, When and Where," *Proceedings of the Michigan Engineering Society* (1908), p. 28.

[98] R. Ben Buckner and Carlisle Madson, "Reprint Preface" (1978) to Wm. A. Burt's *A Key to the Solar Compass* (1881).

[99] The three heirs continuing the lawsuit were John's sons Hiram, age 61, and Alvin, age 59, as well as Austin's son Horace, age 59.

[100] Alan S. Brown, "William Austin Burt: Michigan's Master Surveyor," *Michigan Academy of Science, Arts, and Letters* (1962), XLVII, 274.

[101] Michael H. Adler, *The Writing Machine* (London: Gee and Company Ltd., 1973), p. 159.

[102] Adler, p. 137.

[103] Arthur T. Foulke, *Mr. Typewriter: A Biography of Christopher Latham Sholes* (Boston: Christopher Pub. House, 1975), p. 10.

[104] *Twelfth Census of the U.S., 1900—Manufacturers*, p. 442.

[105] Maury, crippled since 1839 in a stagecoach accident, was recommended by a special Naval Board in 1855 for leave of absence "to promote the efficiency of the navy." He was fighting this decision at the time Burt met with him in 1857, and in 1858, with support from friends and petitions signed by seven state legislatures, Maury was restored to active duty and promoted to rank of commander. In 1861, three days after Virginia withdrew from the Union, Maury resigned from the Navy. Later he served as commander in the Confederate States Navy.

[106] John Burt, Letter to A. D. Bache, Feb. 17, 1857.

[107] A. D. Bache, Letter to John Burt, Feb. 20, 1857.

[108] Wm. J. Young, Letter to A. D. Bache, April 14, 1863. Young had received permission from the Burt family to make the entire instrument or to modify it.

[109] Wm. J. Young, Letter to A. D. Bache, April 21, 1863.

[110] A. D. Bache, Letter to Wm. J. Young, April 28, 1863.

[111] Correspondence of Curtis F. Burt, including a letter from W. M. Dillon, Commander, U.S.N., Naval and Air Attache, March 6, 1940, author's collection.

[112] John Burt, *History of the Solar Compass*, p. 15.

[113] Houghton-Ives field notes, 1863 California-Nevada boundary survey (Books G & H).

[114] Francois D. "Bud" Uzes, author of *Chaining the Land* (A History of Surveying in California), Letter to the author, Sept. 15, 1981. According to Uzes, the operation of the astronomical transit was slower and more cumbersome than the solar compass, but it was accurate to within a few seconds. The solar compass was generally accurate to within one or two minutes.

[115] Butler Ives (1830–1872), Letter to his brother William, Nov. 17, 1867, typewritten copy, California State Library (Sacramento).

[116] John Burt, *History of the Solar Compass*, p. 18. Although the Calumet and Hecla, which operated until 1968, returned over $200 million in dividends, Hulbert did not share in the wealth. He lost his interest in the mine due to poor investments in other Copper Country ventures.

[117] Hiram Burt, *Facts*, pp. 27–28.

[118] Charles E. Smart, *The Makers of Surveying Instruments in America Since 1700* (N.Y.: Regal Art, 1962), pp. 148–149; Commissioner of Patents Report for 1867, p. 1499.

[119] *BLM Manual of Surveying Instructions* (1947), p. 477.

[120] Donald B. Clement, "Evolution of the Solar Transit", *Our Public Lands* (3rd quarter, 1955), pp. 8.

[121] D. K. Parrott, Asst. Commissioner, GLO, Letter to Horace E. Burt, April 29, 1920, cited in Horace E. Burt, *William Austin Burt, Inventor* (Chicago: 1920), pp. 15–16.

[122] Wm. W. Glenn, Chief, Cadastral Survey, BLM office, Portland, Oregon, Letter to the author, June 17, 1983. The Teledyne Gurley Company in Troy, New York, made solar transits for the BLM until 1975.

[123] W.S. Bayley, "Geological Explorations and Literature," *U.S. Geological Survey* (1897), p. 21.

[124] Alm and Trethewey, 1971; Am. Iron Ore Assoc.,1970; cited in Donald A. Brobst (ed.), *United States Mineral Resources*, Geological Survey Paper 820 (Washington: 1973), p. 295.

[125] Paul A. Ott, Jr., "Roughing It in the Frontier Forest," *Michigan Out-of-Doors*, January 1984, p. 33.

[126] By 1983 Minnesota accounted for 70%, while Michigan accounted for 25%, of the nation's output of iron ore, produced in the form of pellets from taconite-type crude ores.[126] The Cleveland-Cliffs Iron Company became the sole owner of the Marquette Iron Mining Partnership with an annual pellet capacity of 2.7 million tons.

[127] *The Traverse Region, Historical and Descriptive* (Chicago: H.R. Page, 1884), p. 123.

[128] A special attraction of Burt Lake is "top flight fishing for perch, pike, and walleyes." According to authors for the Federal Writers Program a few decades ago, "deer, birds, and ruffed grouse are plentiful, and bear and bobcat are occasionally encountered." Federal Writers' Program, *Michigan, A Guide to the Wolverine State* (N.Y.: Oxford Univ. Press, fifth printing, 1949), p. 493.

[129] James H. Mullett, John's son, surveyed the section lines around "Burt's Lake" during the 4th quarter of 1840; his work was resurveyed, however, by Harvey Mellon in 1855.

[130] *Index to Local and Special Acts*, p. 121. See Laws of 1873, Vol. 1, p. 594. Burt Township initially included towns 33, 34, 35, and 36 N, in Ranges 1 E, and 1, 2, and 3 W. Walter Romig, *Michigan Place Names*, n.d., p. 89. For a history of the Burt Lake area see: Helen Boyd Higgins, *Our Burt Lake Story* (Burt Lake Christian Church, 1974).

[131] *A Trip Through the Lakes of North America* (N.Y.: Disturnell, 1857), p. 68.

[132] U.S. Cong., Senate, "Catalogue of Specimens forwarded to Dr. Jackson by John Locke, December, 1847," *Charles Jackson's Reports,* 31-1, U.S. Serial 551, S. Doc. 1, pp. 566–67.

[133] Marie E. Gilchrist (ed), "William Ives' Huron Mountains Survey, 1846," Michigan History Magazine, Dec. 1966, L, fn. 11, 325; C.E. Rademacher, Michigan Dept. of Natural Resources, Letter to the author, Sept. 22, 1983. Mount Burt, or Sugarloaf Mountain, is located in the northeast part of Section 32, T 49 N, R 25 W.

[134] Surveyor Vic Hedman, U.S. Forest Service, Milwaukee, Letter to the author, Aug. 5, 1981.

[135] Donald D. Lappala, Retracement and Evidence of Public Land Surveys (Ironwood, MI: Eastern Region Forest Service, U.S. Dept. of Agriculture, 1974), p. 1.

[136] L.K. Forman, Asst. to the Exec. Dir., The Hall of Fame of Great Americans, Letter to the author, July 26, 1976. The Hall of Fame is affiliated with New York University and the City University of New York.

[137] "Historical Notes", *Michigan History Magazine,* vol. 14 (1930), pp. 339–40.

[138] A newer organization, The National Inventors Hall of Fame, has been created, dedicated "to the individuals who conceived the great technological advances which this nation fosters through its patent system." Each year a selection committee votes on inventor applicants after considering the contribution of the invention to the nations welfare, and the extent to which it promotes the progress of science and useful arts. To submit an application, write: Chairman, Selection Committee, National Inventors Hall of Fame Foundation, Inc., Room 1D01-Crystal Plaza 3, 2021 Jefferson Davis Highway, Arlington, VA 22202.

[139] The two-post historical marker honoring William A. Burt's original homestead will be placed near the boat launching ramp at Stony Creek Lake. This effort, first initiated by the author in 1977, has been coordinated by the Greater Washington Historical Society and the Michigan Historical Commission.

APPENDIX A

THE BURT LINEAGE

RICHARD[1]
b. England, 1580
d. about 1647

RICHARD[2]
b. England, 1629
m. Charity Gallup, 1657
d. Taunton, MA

ABEL[3]
b. Taunton, MA, 1657
m. Grace Andrews, 1685
d. Taunton, MA, 1711

ABEL[4]
b. Taunton, MA, 1692
m. Sarah Briggs, 1722
d. Taunton, MA, 1766

GEORGE[5]
b. Taunton, MA, 1728
m. Susannah Lincoln, 1750
d. Taunton, MA, 1804

ALVIN[6]
b. Taunton, MA, 1761
m. Wealthy Austin, 1779
d. Wales Center, NY, 1841

WILLIAM AUSTIN[7]
b. Petersham, MA, 1792
m. Phebe Cole, 1813
d. Detroit, MI, 1858

———

JOHN[8]
b. Wales Center, NY, 1814
m. Julia Calkins, 1835
d. Detroit, MI, 1886

ALVIN[5]
b. Wales Center, NY, 1816
m. Harriet Amsbury, 1838
d. Cascade, IA, 1846

AUSTIN[8]
b. Wales Center, NY, 1818
m. Lydia Calkins, 1840
d. Detroit, MI, 1894

WELLS[8]
b. Wales Center, NY, 1820
m. Amanda Beaman, 1851
d. Detroit, MI, 1887

WILLIAM[8]
b. Mt. Vernon, MI, 1825
m. Carolyn Curtis, 1847
d. Marquette, MI, 1898

APPENDIX B

THE AUTHOR'S CONNECTION

WILLIAM AUSTIN BURT [7]

|

AUSTIN BURT [8]

|

HORACE ELDON BURT [9]

b.   Mt. Vernon, MI, 1841
m.   Lillie Higgins, 1868
d.   Patchogue, NY, 1932

|

AUSTIN BURT [10]

b.   Detroit, MI, 1870
m.   Mary Ellen Bartlett, 1898
d.   Ontario, CA 1938

RICHARD BARTLETT BURT [11]

b.   Waterloo, IA, 1906
m.   Elaine Lee Alexander, 1930
d.   Santa Ana, CA, 1969

|

JOHN S. BURT [11]
b.   Los Angeles, CA, 1937
m.   Carol Ann Kreiser, 1965

APPENDIX C

THE BURTS' SURVEYING CONTRACTS

William Austin Burt: Appointed U.S. Deputy Survey - November 23, 1833

Contract Date   **Location**

1833   Nov. 25   Mich. Terr. - 12 TS N of Port Huron,
                    $2.75/mi. Assisted by Alvin

1834   Mich. Terr. - Detroit-St. Joseph RR, Detroit-Ypsilanti RR

        Dec. 26   Wisc. Terr. - 12 TS, near Milwaukee
                    Alvin, cook & axeman; Austin, chainman

1835   May   Wisc. Terr.

                    Mich. Terr. - Port Huron-Saginaw City RR

1836   Jan-Jun   Wisc. Terr. - Subdiv. 10 TS

        Sep. 28   Iowa Terr. - Fifth Principal Meridian Line
                    completed Nov. 2 1300 mi. @ $3.50/mi.

1837   May 9   Iowa Terr. - 1341 mi. @ $2.75/mi
                    With Alvin

1839   Oct. 29   Wisc. Terr. - Ext. TS lines

                    Iowa Terr. - 576 mi. @ $4/mi.

1840   Jan. 15   Mich. - Began U.P. surveys 8 1/2 mos. $4.25/mi.

        Feb. 7   Wisc. Terr. - Resurvey work T1N, R15E, 4thPM

        Dec. 3   Mich. U.P. -

1841   Apr. 1   Mich. U.P. - 1111 mi. @ $4.25/mi.

1842   Apr. 11   Mich. - Exam. in Saginaw area

1843   May 25   Iowa Terr. - 550 mi. @ $3.50/mi

1844   Jun. 25   (Houghton Contract): U. P. mineral district

1845   Feb. 26   (Houghton Contract): U.P. mineral district
                               4th correction line to Menominee River.

1846   Sep. 7   Mich. U.P. -Ext. lines of 86 TS, N of
                               Menominee River. Extend 4th correction line
                               from R23W to Menominee River. With John & Austin.

1847   Apr. 27   Mich.-Wisc. boundary

          Apr. 27   Mich. U. P. - TS Lines @ $8/mi.

1848   Mich. U.P. - TS Lines

1849   Apr. 16   Mich. - Exams. in Saginaw area

1850   Jun.   Mich. - (No contract) resurvey- Nicholson district

                          Mich. - Verbal instructions- resurvey
                          Brookfield district

1853   Oct. 5   Mich. - Private Claims near Milk River Point,
                          Lake St. Clair

————

John Burt: Appointed Deputy Surveyor - May 18, 1841

1843   May 27   Mich. - Subdivded north of Saginaw Bay
                          @ $2.75/mi. With John & James H. Mullett

1845   Feb. 26   Mich. - north of Saginaw Bay. 940 mi. @ $4.25/mi.

1846   Sep. 7   Mich. U.P. - with William Austin and Austin
                          Ext. lines of 86 TS.   Extend 4th corr. line.

1847   Apr. 27   Mich. U.P.- (William Austin and Austin's
                          contract) TS Lines, close on Mich.-Wisc.
                          boundary line

1848	Dec. 9	Mich. U.P. - TS 44N, R29 & 30W; TS 47N, R27-31N @ $6.00/mi.
1849	Feb. 23	Mich. U.P.-Subdiv. 7 TS @ $6.00/mi.
	Apr. 26	Mich. U.P. - TS41-42N, R29-30W; TS41-45N, R31W; TS41N, R32W
1850	Apr. 26	Mich. U.P.- mineral district, 10TS W of Escanaba River, N of 4th correction line

———

Alvin Burt: Appointed U.S. Deputy Surveyor on April 27, 1837

1837	May 9	Iowa Terr.- with William Austin @ $2.75/mi.
1838	Aug. 25	Mich.- N. of Saginaw Bay 660 mi. @ $2.75/mi. with Austin T25-26N, R11-16W
	Aug. 25	Mich.- 262 mi. @ $4/mi.
1842	Apr. 12	Wisc. Terr.- 15 TS + frac. TS, Exam. & resurvey of Sprague & Mansfield district N of Wisc. River, E of 4th PM 503 mi. @ $2.75/mi.
1843	Apr. 12	Wisc. Terr.- Private Claims
	May 8	Iowa Terr.- 570 mi. @ $3.50/mi.
	Sep. 30	Iowa Terr- 300 mi. @ $2.75/mi.

———

Austin Burt: Appointed U.S. Deputy surveyor in August 1838

1838	Aug. 25	Mich.- north of Saginaw Bay 600 mi. @ $2.75/mi. with Alvin

1846	Sep. 7	Mich. U.P.- Ext. lines of 86 TS. Extend 4th corr. line. With William Austin and John
1847	Apr. 27	Mich. U. P.- 87 TS bordering on Mich.-Wisc. boundary @ $8/mi.
1848	Dec. 9	Mich. U.P.- (May-Oct, worked without contract) 530 mi @ $5/mi. 10 TS, TS48 & 50, R42-45W; TS50N, R51 & 46W
1850	Apr. 19	Mich.- subdiv. 489 mi. at $4/mi. N of the 3rd correction line
1851	Apr. 10	Mich. U.P.- incl. TS46-48, R46W Several of his party sick, weather poor 179 mi. @ $6/mi.
1852	Apr. 15	Mich.- Correct, retrace, & establish lines of other surv. work. Assisted by William Austin. TS13-17N, R1W; TS7-13N, R2W

––––––

Wells Burt

1848	Dec. 9	Mich. U.P.- Subdivided TS45N, R29 & 30W; TS48N, R27-31W @ $6/mi.
1849	Feb. 26	Mich. U.P.- (bond) 513 mi. @ $6/mi. (7 TS)
	Jun. 29	Mich. U.P.- Subdivided TS 50 & 51, R5W; TS46-51N, R6W; TS45-50N, R7W: Sault Ste. Marie
1850	Apr. 26	Mich. U.P.- Subdivided 5 TS; TS46-50N, R8W 357 mi. @ $4/mi.
1851	Apr. 10	Mich. U.P.- Subdivided 9 TS; TS47 & 48N, RR40W; TS48N, R41W; 179 mi. @ $6/mi.
1852	Jun. 19	Mich. U.P.- TS43-46N, R40W 372 mi. @ $6/mi. To examine & correct surveys W of Whitefish Point, near Grand Marias. Wells sick early in season.

––––––

Appendix C (Continued)

William Burt

1848	Dec. 9	Mich. U.P.- Subdivided 8 TS, 478 mi. @ $5/mi. TS47N, R43-45W; TS49N, R42-46W
1949	Oct. 1	Mich. U.P.- Instructions to examine Higgins district
1850	Apr. 20	Mich.- 9 TS N of the 3rd correction line; TS52N, R1-5W; TS53N, R1 & 2W; R52N, R1W
1852	Apr. 20	Mich.- N of 2nd correction line TS22-24N, R1 & 2W; TS24N, R3W 450 mi. @ $5.50/mi. In his only known reprimand William was told by Surv. General Noble that his work in TS24N, R3W was not performed as well as it could have been.
1854	May 10	Mich. U.P.- Examinations S of 5th correction line, between R21W and R26W
1856	May 21	Minn. Terr.- 133 mi. TS lines @ $13/mi. Included 66 miles of meridian lines
	May 21	Minn. Terr.- TS lines @ $10/mi.
1857	May 29	Minn. Terr.- 223 mi. @ $6.25/mi.

———

APPENDIX D

THE BURTS IN THE IRON MINING BUSINESS

Date	Name	Members	Comments
1853	Lake Superior Iron Company (Burt Mine)	John Burt, Austin, Wells, William	Most productive U.P. iron co. in 19th century
1854	Peninsular Iron Co.	Wm. A., Austin, Wells, John, Hiram, Solon	Marquette, Detroit (Hamtramck)
1866	Marquette & Pacific Rolling Mill	John, William, Alvin	Blast began 1871. Made iron for breakwater
1869	Whetstone Iron Company	William	Marquette
1869	Cannon Iron Co.	Samuel S.	
1872	Erickson Mfg. Company	William, A. Judson	
1872	Huron Bay Iron & Slate Co.	William	Marquette
1873	Burt Mfg. Co.	Austin (Pres.) Horace (Supt.)	Detroit
1873	Union Fuel Co.	John	Marquette
1873	Union Iron Co.	Austin, Wells Lee	Detroit
1874	Carp River Iron Co.	Hiram	Marquette County
1879	Detroit Iron Furnace Co.	Lee (Mgr.)	
1880+	Marquette Furnance Co.	Lee (Mgr.)	Marquette
1880+	Detroit & Marquette Iron Company	John	Marquette
1882	Vulcan Furnace	Lee, Charles	Newbury

Appendix E

BURT'S SOLAR COMPASS

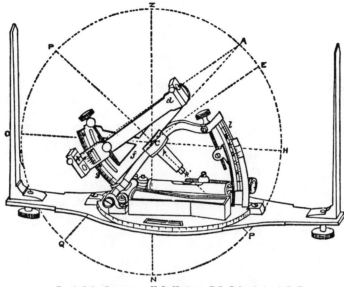

Burt's Solar Compass. H O, Horizon; P P, Polar Axis; A Q, Equator;
A C Z ▬ P C O, Latitude; C A *e* ▬ A C E, South Declination.

Features:

* Bar *f* revolves in the plane of the equator *A Q* about the polar axis *p p'*, carrying the declination arc *g* and bar *d e*. (Bar *f* can be rotated end for end so that bar *d e* corresponds with either north or south declination.

* The sun's rays pass through the lens at the end of bar *d e* and are focused in the center of crosshatched lines etched on a silver plate at *e*.

Operation:

* To locate the true north-south meridian:

    1. Place the solar compass on a tripod and level it.

    2. Set the horizontal plate at 0°00'

    3. Set the latitude *P C O* (from sighting on the sun at noon) on the latitude arc *1*.

    4. Set the declination *A C E* (from ephemeris tables) on the declination acr *g* .

    5. Jointly rotate both the compass about its vertical axis *Z N* and the bar *f* about its polar axis *p p'* until the image of the sun falls in the center of the crosshatched lines of the silver plate. (Additional horizontal lines etched on the silver plate allow the surveyor to correct for refraction. For best results, however, use of the solar compass should be avoided during early morning, midday, or late afternoon.

When the above operations have been completed the sideral time will be shown along the scale of the hour arc (not shown), and the tall sights of the compass will be aligned on the true meridian.

183

Index

Adams, John Quincy, 11
Allen, John, 4, 6, 7
American Industrial League, 146
Anti-Mason Party, 12, 13
Astronomical Compass, Burt's, 92, 93
Astronomical transit, 111, 112, 152, 173

Babbage, Charles, 101, 102,
Bache, D. D., 1331, 132, 150, 158, 171,173
Bailey, John A., 92, 93, 114, 115, 124-126, 165
Baltic, Steamship, 101
Battle Creek, MI., 11
Bay de Noquette and Marquette Railroad, 135, 136
Beaumont, Gustave de, 45
Bingham, Kinsley S., 121, 135, 136, 171
Bonney, 57, 58, 63
Booth, Caleb H., 85, 86, 166
Broadalbin, NY, 3, 4
Brooks, John W., 121
Brooks, Preston S., 131-133
Brown, J. Vernon, 109, 110, 113, 121, 122, 124
Burnham , Hiram, 29, 30, 158
Burt, Abel[3,4] 2, 176
Burt, A Judson[9], 141, 179
Burt & Allen, Millwrights, 4, 7
Burt, Alvin[6], 2-4, 200
Burt, Alvin[8], 6, 9, 26, 28, 29, 33, 38, 47, 55, 57, 65, 67, 82, 160, 161, 163, 177, 178
Burt, Alvin[9], 107, 140, 172, 173, 181
Burt, Austin[9], 6, 9, 23, 28, 47, 66, 67, 75, 77, 79, 82, 95, 117,137, 139, 149, 156, 158,159, 166, 168, 169, 177, 178, 181, 182, 149
Burt, Austin[10], 20,157, 161, 162, 179
Burt, & Bailey, 114, 125, 129, 130,, 137, 170
Burt Block, 140-143, 172
Burt Brothers, 139-143
Burt Freestone Co. 140-143
Burt, George[5], 2, 3
Burt, Hiram[9], 107,140-149, 157, 166, 172
Burt, Horace E.[9], 56, 57, 139, 155, 156, 169, 172, 177
Burt, John[8],5, 9, 28, 29, 47, 50, 52, 53, 55, 57, 65, 67, 75, 76, 79, 80, 81, 82, 92, 95, 96, 107-110, 112, 113, 118-121, 126, 130,131, 133-137,139-154, 157-163, 165, 168, 169,170-172, 173, 177, 179, 181, 182
Burt, J. & W., 125, 126,

Burt Lake, 153, 154, 174
Burt Mine, 135, 136, 142, 143
Burt Mountain, 115, 152, 153, 174
Burt Phebe,4, 5, 9, 10, 14, 16, 20, 26, 27, 30, 47, 50, 56, 57, 85, 86, 137, 139, 157, 158, 160, 161, 162, 165, 166, 167, 168, 171, 172, 177,
Burt, Richard[1], 1, 177
Burt, Richard[2], 1, 2, 177
Burt, Sally, 3, 155, 156,
Burt Township, 153, 154,
Burt & Watson, 125, 126,
Burt, W. Clayton, 172
Burt, Wealthy, 2, 3, 177,
Burt, Wells[8], 6, 47, 57, 67, 77, 79, 80, 81, 82, 84, 95, 96, 109, 110, 115, 117, 118, 137, 139, 149, 177, 183,
Burt, William[8],13, 47, 67, 79, 80, 81, 82, 95, 96, 115, 117, 118, 125, 126, 137, 139, 140, 141, 142, 143, 146, 149, 163, 165, 168, 170, 177,
Burt, William Austin[7], 1-30, 33-39, 43, 45-51, 52-58, 111-113, 115, 118, 124, 125, 127-132, 139, 149-181
Burt, William Austin[9], 140
Butler, Andrew Pickens, 131
Butler, Nathan, 126, 168-171
Butterfield, Justin, 81-86, 94, 98, 104, 105, 111-113, 164-169

California gold, 81-83, 152, 153,
California Surveyor General's seal, 105
Calkins, Elisha, 135, 136, 172
Calkins, Julia, 107, 168
Calkins, Lydia, 168
Calumet and Helca Mines, 152, 173
Canal lock inventiorn, John Burt's, 143, 146,
Canfield, Captain Augustus, 113
Cannon, George115, 156, 157, 164, 165, 169,
Carp River Forge, 107
Carp River Iron Co., 117, 118
CAscade, IA Territory, 56, 57
Cass, Louis, 8, 43, 81, 85, 86, 101, 118, 167
Central Pacific Railroad, 152
Chalets, The Burts', 67, 163
Chapman, Leander, 166
Chicago (IL Territory) 28
Chippewa Indians, 44, 45, 55, 63, 64
Churchman, John III, 91, 92
Cincinnati, OH, 6
Clason, Lewis, 45, 160

Clay, Henry, 12, 109, 110
Clevelnad-Cliffs Iron Co., 79, 80, 172
Cleveland Iron Co., 115, 117
Cobb, Howell, 91
Cole, John & Sarah, 4, 5
Colt, Samuel, 132, 171
Columbia, 123
Compass, Burt's solar38, 39, 47-51, 52, 55, 56, 58-60,
    63,67-72, 76, 77, 79, 80-88, 91, 92, 93, 95-99,
    102-105, 109-112, 114-117, 124-132, 137, 139,
    147, 149, 150-154,
Compass, Burt's variation (solar), 31-34, 37, 39
Compass, magnetic, 24, 25, 26, 29, 31-34, 37, 55, 58,
    60, 63, 64, 65, 80, 85, 91, 92, 98, 104, 105, 112,
    124, 126, 152, 153,
Copper lands of Michigan, 47, 52-55, 60, 63, 66, 79,
    161, 162, 173,
Cram, Captain Thomas, 75, 76,
Crystal Palace, London's, 99, 101, 167

Densmore, James, 149, 150,
Detroit (MI Territory),6, 7, 8, 9, 13, 28,
Detroit Free Press, 13
Detroit Gazette, 9, 13, 14
Detroit & Marquette Iron Co., 143, 146
Detroit & St. Joseph Railroad, 28, 29
Dewey, William, 96
Dexter, Samuel , 12

Elmwood Cemetery (Detroit), 139, 171
Ely, George, 115, 117
Ely, Herman B.,190, 113, 115, 117, 135, 136,
Ely, Samuel, 115, 117, 135, 136, 169,
Equatorial Sextant, Burt's129-131,137, 139, 149, 150,
    152, 171,
Erie Canal, 5, 11
Everett, Philo M., 63-65, 162

Felch, Alpheus, 86-92, 96, 98, 109, 110, 112, 113,
    157, 165, 166, 167, 168, 169,
Fillmore, Millard, 109, 110
Fitzhugh, John, 126, 127
Five Million Dollar Loan, 40
Fletcher, Joseph, 7, 8
Fort Meigs, 6
Fort Wayne, 6

Foster, J. W., 112, 168, 169,
Foster & Whitney, 95, 96, 98, 167, 168,
Franklin Institute, 33, 34, 38, 50,
Freehold, NY, 3

General Land Office (GLO), 28, 50, 51, 52, 56, 75, 76,
    77, 81, 82, 83,
Geological survey of Michigan, 47, 52, 53, 55, 82,
    160, 161, 162,
Geological-linear survey "with reference to mines &
    minerals", 57, 65, 66, 67, 77, 79, 161, 162, 163,
Gibbs, W., 77, 79
Glidden, Carlos, 149, 150
Gogebic iron range, 162, 163
Goodyear, Charles, 103, 104
Grace Furnace,143, 146
Grant & Crosman, 125, 126, 170
Grant, Ulysses S., 143, 146
Graveraet, Robert J., 107, 109
Greater Washington Historical Society, 175
Greeley, Horace, 167, 168
Gurley, W. & L. E., 79. 80, 152, 153, 170, 171

Haines Ezekiel S., 38, 39, 45, 47, 50, 51, 52, 53, 54,
    55, 16, 161,
HAll of Fame of Great Americans, 175
Harvey, Charles T., 114, 115, 117, 123, 124,
Herschel, Sir John, 102, 103
Higgins, Hiram, 66
Higgins Lake, 98, 164, 165
Higgins, Sylvester, 66, 77, 82,
Hough, Benjamin, 6
Houghton, Douglass, 47, 52-58, 60, 65, 66, 79,
    160-163
Houghton, Jacob, 57, 58, 60, 160, 162, 163
Houghton, MI, 66
Hubbard, Bela, 66, 162, 163, 165
Hulbert, Edwin J., 152, 153,
Huntingtonl, Elisha, 51, 55,

Illinois, 57, 123, 124, 130,
Independence, 78
Instructions, General Surveying, 27, 29, 31, 47, 55,
    84, 96,
Instructions, Manual of Surveying, 125, 126, 127, 132,
    170, 171,
Instructions, 1850 Michigan General, 95, 96, 157, 158

Instructions, 1851 Oregon General, 103, 104, 112, 113, 125, 126
Iowa-Minnesota boundary survey, 111, 112, 168, 169,
Iowa-Missouri boundary survey, 96
Iowa surveys, 35-37
Iron Mountain Railroad, 109, 114, 115, 117, 121, 135, 136.
Iron ore, 29, 30, 46, 54, 56, 60-67, 79, 109, 110, 161, 162, 172, 174
Ives, Butler, 152, 173
Ives, William, 57, 58, 60, 1104, 105, 161, 162, 173, 174,

Jackson, Charles T., 77, 162, 163, 173, 174,
Jackson Iron Co.,60, 63, 64, 65, 109, 126, 162, 168
Jackson Mining Co., 63, 64, 107
John Burt House, 139, 140, 171,
Jones, George W., 69, 163
Juneau, Solomon, 28, 29,

Kansas-Nebraska bill, 118
Keitt, Lawrence M., 131`, 132, 133
Key to the Solar Compass, 127, 129, 172, 173
King, James, 57, 58
King, Samuel, 105

Lac Vieux Desert, 75, 76
Lake Superior Iron Co.,107, 109, 115, 117, 118, 135, 136, 142, 143, 146, 154, 172
Lake Superior Journal, 123
Lake Superior Mining Journal,, 123, 124
Lake Superior News, 66, 67
Land Ordinance of 1785, 23-26
La Pointe, Treaty of, 55
Lawton, Charles D., 152
Lincoln, Abraham, 29, 81, 91, 158, 164, 165, 166
Locke, John, 96, 98, 167, 173, 174
London's 1851 Great Exposition, 129-132, 167
Lyon, Lucius, 7, 9, 27, 28, 29, 35, 37, 38, 43, 44, 52, 53, 56, 65, 66, 70, 71, 72, 75, 76, 79, 80, 81, 82, 83, 84, 87, 88, 93, 94, 95, 96, 155-166
Lyon, Orson, 35
Lytle, Robert, 29, 30, 44, 45, 158, 159, 160

Mackinaw, MI, 57, 58
Mail service, U. S., 6, 21, 23, 123, 124

Marji-Gesick, Chief, 63, 64, 162
Marquette Brownstone Co., 142
Marquette breakwater, 141
Marquette County Historical Society, 139
Marquette Furnace Co., 143, 146
Marquette iron range, 57-67, 79, 107, 109, 152, 153, 154, 161, 162
Marquette, MI, 121, 125, 135, 136, 139-147, 171, 172

Marquette & Pacific Rolling Mill Co., 140-143
Marsh, James M, 111, 112
Martin, E. C., 75, 76,147, 149
Martin, Morgan, 39, 40, 158, 159
Mason, Stevens T., 39, 40
Masonic Order, 5, 12, 13
Maury, Matthew F., 130, 13`, 149, 150, 171, 173
McCormick, Cyrus, 103, 104
McKnight, Sheldon, 109, 110
McLeod, W. Norman, 57, 58
Mellen Harvey, , 57, 58
Menominee iron range, 162, 163
Meridian, Fifth Principal, 35, 37
Meridian, Fourth Principal, 80
Meridian,Michigan, 6, 45, 47
Mesabi iron range, 153, 154
Michigan Engineering Society, 147, 149
Michigan Farmer, 92, 93, 124, 125, 165, 166
Michigan Historical Commission, 175
Michigan Legislative Council, 11, 13, 14
Michigan's "internal improvements", 40
Michigan Society Registered Land Surveyors, 47
Michigan statehood, 33, 43, 44, 45
Michigan's State Legislature of 1853, 113, 114
Michigan State Line Historical Site, 75, 76, 164
Michigan U.S.S., 60, 162
Michigan-Wisconsin boundary survey, 57, 75-79, 164
Miller, Hugh, 103
Mill, Henry, 21, 22
Mills, Construction of, 5, 6, 7, 9, 11, 12, 23
Milwaukee survey, 28, 29, 33
Missouri Compromise, 118
Monitor and the Merrimac, 150
Monticello, 109
Mt. Vernon MI., 7, 23, 28, 30, 67, 79, 91, 107, 139,
Mullett, James H., 7, 11, 12, 23, 29, 34, 50, 55, 56, 64, 65, 79, 82, 83, 94, 95, 96, 149, 153, 154, 156, 158, 162, 163, 166,
Mullett, John H., 149, 160, 161
Mullett Lake, 153, 154

National Inventors Hall of Fame, 175
Negaunee, MI, 63, 64, 162, 163
Neitz, Leonard, 124, 170
New York Commercial Advertiser, 14
New York Herald, 101
Nicholson, Henry, 82, 87, 94, 95, 96, 164, 165,
Noble, Charles, 84, 94, 95, 165-169, 170,
Nolan, Lewis, 63

Ontonagon bolder, 53, 54

Parke, Hervey, 7, 8, 9, 27-29, 35, 37, 155-158,
Patten, Richard,
Patents received by William A. Burt: 14, 16, 18, 33, 34
Patents received by John Burt: 146
Paxton, Sir Joseph, 101
Peninsular Iron Co., 115, 118, 142, 143
Petersham, MA, 2, 3
Pilkington, Margaret,, 115, 117, 169
Poe Lock, 146
Poole, William C., 96, 165, 166,
Pratt, John, 149, 150
Preston, John, 98, 111, 167, 168
Prince Albert, 100, 101, 104

Queen Victoria, 100, 101, 104

Republic Mine, 79
Revolutionary War, 2, 4
Risdon, Orange, 81, 82

Sam Ward, 78
Sargent, George B., 111, 112, 168, 169
Sault Ste. Marie, MI.,47, 56, 57, 63, 75, 76, 94, 95, 107-110, 113, 114, 123, 124
Sax & Fox Indians, 55
Schmolz, William, 152, 153
Scott (John) Legacy Medal, 158, 159
Searlles, Zelotes, 109, 155, 156
Sheldon, John, 13-21
Shoenberger, Peter, 109, 110, 123, 124, 170
Sholes, Christopher Latham, 149, 157
Simpson, Thomas, 118, 165
Smith, Benjamin H. 152, 153

Smith, Moses, 91, 165, 166
Smith, Truman, 117
Soo Canal, 40, 54, 55, 109, 110, 112, 113-115, 121, 123, 124, 130, 133-135, 146, 168, 169, 171, 172, 173,
Soule, S. W., 149, 150
Spalding, Cyrus, 20, 21, 157
St. Louis, MO Territory, 6
St. Mary's Falls Ship Canal Company, 114
Stockton, John, 156
Stony Creek (MI), 7, 9, 11
Stony Creek Lake, 175
Stuart, Charles, 118, 119, 131
Sumner, Charles, 131-133
Sun dial, Burt's, 51
Superior, 7, 8, 9
Survey inspector, 85

Talcott, Captain Andrew, 111, 112
Taunton, MA, 1, 2, 155, 156
Taylor, John, 57, 58, 63
Taylor, Zachary, 81, 82
Teal Lake, 57, 58, 60
Tiffin, Edward, 6, 7, 11, 155, 156
Tocqueville, Alexis de, 45
Toledo strip, 43, 44
Transit, Solar, 153, 173
Transit, Surveyors, 33, 71, 72, 158
Traverse City, MI., 94
Typographer, Burt's (typewriter)., 13-22, 149, 150, 157, 172, 173,

Union Fuel Co., 143, 146
U. S. Department of the Interior, 81

Van Buren, Martin, 14, 81
Vanderbilt, MI., 95, 96
Variation of the compass, 25, 26, 27, 29, 31, 50, 58, 60, 95. 96, 112, 126, 127, 130, 150

Walbridge, David, 131-133, 168, 169
Wales Center, N.Y.,4-7, 57,
Walker, R., 96
Walk in the Water, 7
Ward, Eber Brock, 109, 110, 114, 168
Ware, Henry, 38

War of 1812, 5
Washington, George, 2, 12
Washington, Treaty of, 45
Webster, Kimball, 167, 168
Weitzel Lock, 146
Western Emigrant, 12
Whig Review, 100
Whitcomb, James, 45, 160
Whitcomb, Rasselas Rince, 39, 40, 56, 159, 161,
White, Peter, 109, 110, 115, 121, 123, 124, 140-142,
    162, 168, 169, 170,
Whittlesey, Charles, 93, 98, 165-167
Williams, Samuel, 11, 162, 163
Williams, Micajah T., 23, 26, 28, 29, 30, 157, 158,
Wilson, John, 28, 84, 85, 91, 92, 94, 111, 112, 113,
    126
Wiltse, Harry, 77
Wing, Austin, 12, 160
Winnebago Indian cession, 46
Witness Tree, 166
Woodbridge, William, 7, 40, 52, 54, 160, 161

Young, Alfred, 148, 171, 172
Young, Richard M., 79
Young, William C., 124
Young, William J., 33, 34, 37-39, 47, 50-52, 56. 65,
71, 72, 79, 80, 92, 93, 98, 99, 103, 104, 114, 129,
130, 131, 139, 150, 158-171